W0017544

Increasing Resilience to Climate Change in the Agricultural Sector of the Middle East:

The Cases of Jordan and Lebanon

Dorte Verner, David R. Lee, Maximillian Ashwill, and Robert Wilby

THE WORLD BANK
Washington, D.C.

ISBN (paper): 978-0-8213-9844-9
ISBN (electronic): 978-0-8213-9845-6
DOI: 10.1596/978-0-8213-9844-9

Cover photo credit: © Dorte Verner, used with permission.

Library of Congress Cataloging-in-Publication Data.

Increasing resilience to climate change in the agricultural sector of the Middle East : the cases of Jordan and Lebanon / Sustainable Development Department, Middle East and North Africa Region.
    p. cm.
Includes bibliographical references and index.
ISBN 978-0-8213-9844-9 (alk. paper)
1. Crops and climate—Lebanon.   2. Crops and climate—Jordan.   3. Climatic changes—Government policy—Lebanon.   4. Climatic changes—Government policy—Jordan.   5. Agriculture and state—Lebanon.   6. Agriculture and state—Jordan.   I. World Bank. Middle East and North Africa Region. Sustainable Development.
  S600.64.L4I53 2013
  363.738'740956—dc23

2013001567

Contents

Acknowledgments

The team that compiled this report was managed by Dorte Verner. The report was drafted by David R. Lee, Maximillian Ashwill, Robert Wilby, and Dorte Verner. Additional written background notes and input were provided by Thea Whitman (Agriculture sector information), Shadi Atallah (Agriculture sector information), and Sanne Tikjøb (Country information). The workshops for this task were held in Jordan and Lebanon and were organized by Johanne Holten and David Lee in collaboration with Emad Adly and the country teams in Jordan and Lebanon.

The World Bank is grateful for the endorsement of the proposed actions provided in this report by Agricultural Minister Hussein Al Hajj Hassan of Lebanon and for the ideas and guidance provided by Vahakn Kabakian and Lea Kai Aboujaoudé, of the Ministry of Environment in Lebanon.

The World Bank is grateful for the ideas, comments, and contributions provided by the Ministry of Agriculture and the Ministry of Environment of Lebanon, and the Ministry of Agriculture of Jordan.

The World Bank gratefully acknowledges Randa Massaad of the Lebanese Agricultural Research Institute (LARI) for her leadership in producing the country report for Lebanon and also thanks the Lebanese team members who provided inputs for the report: Ihab Jomaa, Elia Choueiri, Ali Chehade, Ahmed Bitar, Charbel Hobeika, and Machatel Loubnan. The World Bank also wishes to thank Ghada Al-Naber and Muna Saba, from Jordan's National Center for Agricultural Research and Extension (NCARE), for their leadership in producing the country report for Jordan, and thanks team members Faisal Awawdeh, Mustafa Rawashde, Hesham Athamneh, Naem Mazahreh, and Hassan Asof, for their input.

Additionally, the team also gratefully acknowledges support and guidance from Luis Constantino, Svetlana Edmaedes, Hedi Larbi, Laszlo Lovei, Junaid Kamal Ahmad, Hoonae Kim, Pilar Maisterra, Stefano Mocci, and Maurice Saade, and Janette Uhlmann. The team is also grateful for Peer Reviewer comments from Willem Janssen, Eija Pehu and William Sutton. The team also acknowledges support from Marie Francoise How Yew Kin, Dylan Murray, Indra Raja, and Sanne Tikjøb. The team is very grateful to Eliot Cohen for excellent and continuous support and advice on photovisuals for this report, including photo selection, photoediting, etc. Finally, but not least, the team would like to thank the European Commission and Italian government for funding of this task through the MENA Multi-donor Trust Fund.

Executive Summary

Photograph by Dorte Verner

Actions to Increase Resilience Are Needed in Jordan and Lebanon

It is getting hotter and dryer in the Arab countries. This changing climate impacts the lives and livelihoods of rural people in Jordan and Lebanon. This report presents policy directions to increase resilience in the agricultural sector to climate variability and change.

This report aims to assist Jordan and Lebanon in understanding the specific challenges and opportunities posed by climate change in the agricultural sector. The report presents local-level priorities, informed by stakeholder input, to build agricultural resilience in both countries. The objectives of this study were three-fold: (1) to improve the understanding of climate change projections and impacts on rural communities and livelihoods in selected regions of Jordan and Lebanon, specifically the Jordan River Valley and Lebanon's Bekaa Valley; (2) to engage local communities, farmers, local experts, and local and national government representatives in a participatory fashion in helping craft agricultural adaptation options to climate change; and (3) to develop local and regional climate change action plans that formulate recommendations for investment strategies and strategic interventions in local agricultural systems. Therefore, this report may serve as the analytical underpinning for ongoing discussions taking place within the governments of Jordan and Lebanon, including the Ministries of Agriculture and Environment, on how to best move forward in building agricultural resilience to climate change.

An increasing sense of urgency characterizes the global dialogue on climate change. The landmark Fourth Assessment Report of the Intergovernmental Panel on Climate Change (IPCC 2007a) concludes definitively that "[w]arming of the climate system is unequivocal" and goes on to elaborate the expected pervasive effects of future climatic changes. These include rising air and ocean temperatures, extensive melting of snow and ice, rising sea levels and regional impacts ranging from modest to severe. The World Bank's World Development Report 2010 argues that rising greenhouse gases have fundamentally "transformed the relationship between people and the environment," and that, "left unmanaged, climate change will reverse development progress and compromise the well-being of current and future generations." The implications for people in Arab countries are expected to be particularly severe, given their greater exposure to natural hazards, environmental risks, sometimes-precarious livelihoods, and the limited economic and institutional buffers to moderate the negative impacts of climate change. This is clearly spelled out in the report, "Adaptation to a Changing Climate in the Arab Countries" (Verner 2012).

The climate challenges confronting development in the Middle East are particularly stark. This region, and specifically its rural people, face what might be called a "triple threat" from climate change. First, the Middle East is already one of the driest and most water-scarce regions of the world (World Bank 2011a) and faces severe challenges posed by high temperatures and limited water supplies. Higher temperatures, even with no changes in precipitation, can be expected to increase soil moisture evaporation and thus pose the risk of increased soil degradation and related effects (IPCC 2007b). Second, although IPCC (2007b) projections of future changes in precipitation trends in the Eastern Mediterranean region are uncertain, there is a "strong consensus" that annual precipitation totals will fall, and it is estimated that precipitation could be reduced by 10–20 percent (chapter 2). Such declines would further exacerbate the already precarious water scarcity facing the inhabitants of much of the region, who face severe water constraints for both human and agricultural use. Third, the rural economies of the region, as well as many rural people, are highly dependent on agriculture.

In Lebanon, agriculture accounted for roughly 6.1 percent of gross domestic product (GDP) in 2008, employed an estimated 20 percent of the labor force, and has contributed positively to GDP growth in recent years (IMF 2010; MOA (Lebanon) 2004). In Jordan, the agricultural sector represented nearly 3 percent of GDP in 2010, but constitutes the main livelihood for 20 percent of the population and employs 9.8 percent of its economically active labor force. Yet the agricultural sector is highly vulnerable to climate change. The sources of this vulnerability are many, including reduced yields, the likelihood of crop failures due to severe weather events (droughts, floods, extremes in temperature), the increased incidence of crop pests and disease, and other factors. In the Middle East, a recent study by the International Food Policy Research Institute suggests that climate change impacts on crop yields (to the year 2050) will be particularly severe: a 30 percent decline for rice, 47 percent decline for maize, and 20 percent decline for wheat (Nelson et al. 2009).

For all these reasons—the dependence of their livelihoods on climate and natural resources, their vulnerability to rising temperatures and water scarcity, and the increased likelihood of future drought and severe weather events— farmers and rural households are already on the "front lines" of dealing with and adapting to climate change. Farmers constantly make autonomous adaptions in the face of ongoing changes to agroecosystems, not just climate change. These autonomous adaptation strategies include changing crop selection, allocating land between alternative crop and livestock uses, and adjusting planting and harvesting dates. The rural poor are especially vulnerable to climate change impacts because of their disproportionate dependence on agriculture, lower ability to adapt, and frequent lack of support systems to facilitate adaptation (World Bank 2008). Grasslands, livestock and water resources in the region—and the rural households dependent on these resources—are expected to be particularly vulnerable to climate change (IPCC 2007a). Beyond the level of individual farms and households, the impacts of climate change on local social structures and agroecosystems in the Middle East and elsewhere will be significant. Land degradation may result from decreased soil moisture and overgrazing. In addition to the impacts on productive capacity, climate change may also trigger major effects on health and food security in the region. Heat stress and increases in vector-borne and water-borne disease are expected to occur. Other impacts include decreases in caloric availability of up to 500 calories per person per day and in an increase of one million malnourished children in the North Africa and Middle East compared to a "no climate change" scenario (Nelson et al. 2009).

To deal with these growing constraints, effective adaptation is critical. In addition to autonomous adaptations, agricultural climate adaptation can be planned (FAO, 2007). Planned adaptations are deliberate long-term response strategies and policy interventions developed in response to the recognition of changed conditions and are aimed at improving the resilience of agroecosystems. These adaptations can include the spontaneous adaptations mentioned above but can also go well beyond these to encompass research and technology development, investments in irrigation systems, genetic improvements and crop varietal development, and institutional changes in intellectual property rights, land-use and agricultural policy. Because many of the impacts of climate change on agriculture are predictable—qualitatively if not quantitatively—planned adaptation is widely considered to be a prudent and efficient response, and many organizations have urged that financial resources be devoted to investing in agricultural adaptations (Nelson et al. 2009; World Bank 2007). The public sector has a significant role to play in making these investments and providing an institutional and policy framework that facilitates adaptation strategies. Since many of these investments are part of good agricultural development policy anyway, they can yield "double dividends," in that they not only represent valuable and appropriate measures to foster agricultural development but they also contribute to agroecosystem resilience to sources of climate risk (Padgham 2009). Ultimately, the effectiveness of adaptation strategies depends on how well they deal with the diverse local conditions faced by farm households and rural communities (Agrawal, McSweeney, and Perrin 2008; World Bank 2009).

A Participatory Process Can Be Used to Build Climate Resilience

The challenge in developing agricultural adaptation strategies to locally relevant solutions frequently entails addressing two interrelated goals. These are, first, decreasing the vulnerability of local communities to climate variability and change, and second, increasing the resilience of natural and social systems in the face of climate change (World Bank 2011b). Developing solutions to address these goals, however, quickly encounters two major challenges. As agriculture is a natural resource-based industry and one heavily dependent on varying local conditions in climate, soils and other physical and community characteristics, the first challenge in agricultural adaptation is to identify appropriate solutions to climatic change that address specific local needs. The second challenge is one of prioritizing potential response options in a manner that effectively addresses climate change while making the best use of the inevitably constrained resources.

In order to develop the local-level priorities for agricultural adaptation to climate change in Jordan and Lebanon, a four-step priority-setting approach was used here. This approach, which has been successfully used elsewhere (World Bank 2009), uses science-based climate projections and a participatory multicriteria scoring method to identify and prioritize alternative strategies for agricultural climate adaptations. The priority-setting process was built around a series of workshops and meetings with stakeholders and policy makers (described in detail in the "Methodology" section in chapter 3). Two country teams led the activities in each country, which included the preparation of the action plans. In Jordan, the country team was led by the National Center for Agricultural Research and Extension (NCARE). In Lebanon, the Lebanese Agricultural

Research Institute (LARI) undertook the central organizing role for the study's work. Scientists, researchers, and officials from national government institutions, universities, and other organizations played active roles as presenters and advisers at the workshops. Other participants included farmers and representatives of farmer organizations, extension specialists, university and research personnel, representatives from nongovernmental organizations (NGOs) and local governments, journalists, and members of international development and research organizations. World Bank staff and consultants also played an important supporting role in the workshops.

It is important to highlight several aspects of this study. First, the focus is on climate change adaptation, not mitigation. As mentioned below, the two countries that are included in this study, Jordan and Lebanon, are minimal contributors to total global greenhouse gas (GHG) emissions. While efforts are underway in each country to curtail emissions, and these efforts should be supported and expanded, the primary challenges in the agricultural sector with respect to climate change in both countries are ones of adaptation. As suggested in other contexts, many "climate compatible" adaptation options may simultaneously reduce GHG emissions and promote mitigation (Mitchell and Maxwell 2010). However, the main thrust of this project is identifying and prioritizing local agricultural adaptation strategies and response options in the two regions.

Second, the approach followed in this study emphasizes the participation of diverse local stakeholders in formulating response options that adapt to climate change impacts on agriculture. This "bottom-up" approach is an important complement to the frequent "top-down" approaches taken in many climate change strategies and national action plans. To some extent, such top-down approaches are inevitable in promoting mitigation strategies, which necessarily involve tradeoffs among sectors and regions, and where establishing national goals and policies to achieve those goals is required. Yet, in formulating adaptation strategies, the local focus is indispensable, so any methodology or approach must involve the input of local stakeholders in a bottom-up fashion.

Third, it should be emphasized that the agricultural adaptation response options identified and prioritized in this study, and the local and regional action plans which build on those response options, are only a first step in generating the needed investments, policy changes and changes in private sector decision making. As in previous applications of this priority-setting methodology (see World Bank 2009), the overall objective is to identify priority interventions, based on local stakeholder inputs, that will facilitate building climate resilience in agriculture. The objective was not to conduct a comprehensive economic feasibility analysis or cost-benefit analysis of all of the response options considered. The priority-setting approach is not a substitute for rigorous economic assessment of possible investments and interventions; rather, the two approaches are complementary. Further detailed technical and economic analyses are a necessary second step, particularly in cases where major public investments are involved, such as in the construction of new irrigation infrastructure or establishing climate early warning systems.

Jordan and Lebanon Are Getting Hotter, Drier and Experiencing More Climate Variability

The evaluation of climate change and variability in Jordan and Lebanon, carried out for chapter 2, is based on a range of primary and secondary data sources held by national and international agencies. These include conventional meteorological records, derived climate indices, and satellite products. This data supports the view that the region has experienced rising air temperatures since the 1970s and shrinking snow cover. There have also been local reductions in rainfall since the 1960s, set against a background of large inter-annual variability linked to the behavior of the North Atlantic Oscillation, which controls the strength and direction of westerly winds and storm tracks across the North Atlantic. Trends in extreme events are more problematic to assess, but there are indications that the frequency of hot nights has risen and heavy rainfall events have become more frequent and intense. Average river flows and groundwater levels have generally declined but it is difficult to disentangle climate-driven trends from those caused by growth in water abstractions over the same period.

Regional climate change projections suggest that these trends are set to continue over coming decades, potentially exacerbating the imbalance between water supply and demand. There is a strong consensus amongst climate models that the region will warm more rapidly than the global average and that annual precipitation totals will fall. By the 2050s mean temperatures could increase by 1.3–2.7°C and precipitation could be reduced by 10–20 percent. Future changes in rainfall will reflect the interplay between a possible northward shift in the Mediterranean storm track, counteracted by rising sea surface temperatures and more frequent polar intrusions. Weaker and/or less frequent west-east airflows would tend to reduce orographic rainfall in the region's mountain ranges.

A statistical downscaling model was used to evaluate site-specific outcomes of the changes in atmospheric conditions. The model was calibrated and tested for 10 locations representing diverse physiographic conditions and agro-ecological zones. Daily mean temperatures and precipitation totals were downscaled from the UK Met Office HadCM3 climate model under two emissions scenarios (SRES A2 and B2) for the period 1961–2099. Downscaled changes in seasonal mean temperatures and precipitation amounts bracket the findings of earlier studies.

Overall, there is a tendency for more rapid warming the greater the elevation and distance from the coast. Warming is most pronounced in spring at coastal sites and for summer at locations inland. The largest reductions to annual rainfall are found for sites in the coastal zone, and within the Bekaa Valley, where changes could be in the range of 10–30 percent by the 2050s and 20–50 percent by the 2080s. Discerning changes in precipitation for the 2020s is generally problematic due to climate model and downscaling biases, combined with large natural variability in annual totals.

A major strength of the statistical downscaling is the ability to derive indices of change that are meaningful to planners such as the likelihood of dry-years, winter growing degree days, or dry season duration. A case study for Amman (Jordan) showed that the chance of a dry year (<200 millimeters) was historically once in three years, but could become as likely as not (that is a 50 percent chance) by the late 2020s. Extension of the dry-season duration by ~30 days by the 2050s could limit the length of the grazing season.

The focus of chapter 2 is on changes in temperature and precipitation over the study sites. Moreover, sea level rise could increase the risk of saline intrusion to coastal aquifers further limiting the resources available to irrigators and urban areas such as Beirut (Lebanon). More intense rainfall could also limit the effectiveness of groundwater recharge. Preliminary research undertaken elsewhere indicates that the future water security of the region will hinge, in part, on changes in snowpack and residual river-flows in the Euphrates, and on negotiated water-sharing with Israel. Likewise, future food security will depend on the ability of domestic and global producers to adapt to changing conditions. Therefore, any sector planning will need to take account of potential impacts of climate change on the environment, water politics and agro-economics of the wider regional neighborhood.

The regional climate change projections and uncertainties described in chapter 2 are a cause for concern given existing water supply deficits and a legacy of overexploited freshwater stocks. However, in the short and medium-term, population and economic growth are more important drivers of the water deficit than climate change. Possible exceptions include situations where a tipping point (such as the limit to rainfed agriculture by ~200 millimeters per year) is being approached. Under more rapid climate change the threshold would be reached sooner, so uncertainty in the climate modeling can translate into uncertainty about the time-scale available for anticipatory adaptation. Further research is underway to better characterize the uncertainty in regional projections by downscaling from more climate models and emissions scenarios.

Developing Action Plans Is a Key Step in Addressing Climate Change

Chapter 3 describes the priority-setting methodology and its results. At the conclusion of this process, stakeholders in each study region ranked a number of priority response options that should be taken to increase agricultural resilience to climate change. These response options comprise the proposed action plans developed for each country and are summarized in order of highest-to-lowest priority in table O.1 on next page.

Table O.1 Summary of Action Plans for Lebanon and Jordan

Lebanon	Jordan
Adopt new irrigation technologies	Improve farm production systems and productivity
Launch project to construct small- and medium-scale water harvesting reservoirs	
	Improve on-farm water use efficiency and integrated water resources management
Integrate the production management of pests, diseases and plant disorders under climate change	Improve livestock and rangeland systems
	Build national capacity for climate change adaptation
Produce and distribute plant materials adapted to climate change	Reduce risks of agricultural pests and diseases
Increase capacity for climate change adaptation	Reinforce early warning system for drought
Evaluate and maintain the genetic diversity of wild species and local varieties adapted to climatic change	Reform land-use laws and implement sustainable land-use
	Activate an agricultural risk management fund

Source: World Bank data.

In Jordan and Lebanon, there was a high degree of commonality in terms of the prioritized response options from each country's action plan. As a consequence, this strengthens the argument that these are urgent actions to be taken, generally, for both countries. In both action plans, addressing water and irrigation-related constraints ranked at the top in terms of priorities. In Lebanon, the two top-ranked response options were explicitly related to irrigation and water management: promoting the adoption of new irrigation technologies through demonstration projects related to drip irrigation—with the potential to greatly economize on water use—and fertigation technologies; and establishing a pilot program for the construction of small- and medium-scale water harvesting reservoirs to provide water storage and supplementary irrigation. The top-ranked response option in Jordan was increasing farm production and efficiency, but many of the proposed activities under this response option relate to improving the efficiency of water use. These activities include: avoiding agricultural expansion into fragile rainfed lands; introducing drought-tolerant crop varieties; identifying alternative cropping patterns that recognize water-related constraints; and promoting conservation agriculture in dry areas. The second-ranked response option in Jordan—increasing water efficiency—was explicitly related to water management. This option encompasses a variety of approaches to improve on-farm water use efficiency and the integrated management of water resources. This includes rainfall harvesting, assessing the feasibility of using treated wastewater and brackish water for irrigation, and developing a system for strict monitoring of groundwater to prevent overexploitation. As expected, water-related constraints were the dominant concern of most local stakeholders participating in the project workshops and were generally the most highly prioritized.

There were other priority response options shared across Lebanon and Jordan, beyond those that are water-related. These included the development of crop varieties that are tolerant of drought, heat and other expected climatic changes (in Jordan, subsumed under "Increasing farm production and efficiency"); a focus on integrated pest management (IPM); and the development of improved local capacity to adapt to climate-related impacts on agriculture. "Improved local capacity" was generally interpreted to incorporate different types of capacity building among various stakeholders. Some examples of these different types of capacity building include: improved understanding of climate change, information provision and training for farmers; an enhanced general recognition of climate change problems among the general public; and importantly—since this need is often under-recognized in national research systems—improving the institutional capacity of researchers and the research system itself to deal with climate change-oriented problems.

In Jordan, the response options had a stronger policy orientation than in Lebanon. In addition to building capacity, these included: the development of a national climate change strategy; the implementation of new land-use laws to foster land-use changes; and the putting into practice of the already approved Agricultural Risk Management Fund (agricultural insurance scheme). The response options in Lebanon assumed more of a research orientation, including research-based initiatives on irrigation technology, crop pests and diseases, and crop varietal development and associated research in applied crop genetics. As expected, many of the response options in Lebanon involved a key role for the Lebanese Agricultural Research Institute (LARI), while in Jordan the National Center for Agricultural Research and Extension (NCARE) figured prominently in the proposed initiatives.

Overall the action plans' priority response options are highly consistent with, and reinforce the importance of, the strategic priorities identified in other research. For example, the World Bank's Middle East and North Africa Region (World Bank 2012a) has identified three broad areas for strategic partnerships between the Bank and its counterparts to address challenges related to improved climate change adaptation. These include infrastructure investment, knowledge strengthening and policy reform. Most of the abovementioned priority response options fall generally within these categories. Two of the Food and Agriculture Organization's regional priorities for responding to climate change impacts in the Middle East region include "improving national and regional capacities to cope with adverse impacts of climate change," and "identifying practices for adaptation and mitigation of climate change impacts" (FAO 2011). Both of these priority areas are directly addressed by the response options related to capacity building. For Lebanon specifically, a recent World Bank review (Lampietti 2010) highlights three challenges facing Lebanese agriculture—infrastructure (irrigation, etc.), water management and urbanization—and identifies a number of specific strategies to address these challenges; these too are addressed by many of the proposed response options. Finally, the World Bank's new Flagship Report (Verner 2012) on climate change in Arab countries identifies a number of strategies and investments to address climate change in the region. These include: technological innovations; institutional strengthening; improved research tools; farm income diversification; and policy reforms. Each of these was addressed in the specific response options identified and prioritized by local stakeholders in this study.

The priority-setting methodology followed in chapter 3 of this study proved to be a practical and viable approach on several levels. The approach was a practical and transparent way to involve local stakeholders in the identification of response options that address climate change adaptation needs in agriculture. It allowed for a prioritization of those options under conditions of limited resources. It also led to the drafting of action plans that could be easily communicable at the policy-making level. This "bottom-up" approach is centered around the input of local stakeholders from the outset and thus assures that the response options that are prioritized address local needs as viewed by farmers, researchers, extensionists, and others involved at the field-level. There were no discernible problems in eliciting the input of farmers and other local stakeholders on the subject of climate change. Farmers' yields and incomes are directly tied to the natural resource base on which they depend, so they are acutely aware of changes in that resource base—particularly regarding often-limiting water resources—and were eager to share their views and opinions in the workshops organized for this study.

In general, the development of the action plans served their function as a necessary "first step." However, in order to successfully achieve needed investments, interventions and policies that can serve to locally address climate change impacts on agriculture, this information must be shared with policy makers. Furthermore, these policy makers must be willing to act. In Jordan and Lebanon, the success of this varied. At the time of publication, the extent to which the Ministry of Agriculture in Jordan would consider the action plan recommendations was unclear. A major reason for this was continued uncertainty regarding personnel changes in the King's cabinet. In Lebanon, by contrast, the action plan was presented directly to the Minister of Agriculture, who agreed to implement some or all of the priorities. Many details are to still be decided related to the structure, scope and budget of a proposed intervention, but the potential use of this methodology has proven its worth.

The methodology was successful in breaking down the collection of information into a series of "manageable parts." The sequence of steps in the priority-setting process, which are built around two workshops and a final decision meeting, permits this breakdown. As noted elsewhere (World Bank 2009), this has several advantages. First, it facilitates the distinction between—and reduces confusion among—the identification of response options versus evaluation criteria. By explicitly identifying, assessing and weighting the evaluation criteria, it becomes easier to distinguish between "decision options" and "criteria" for evaluation that sometimes become conflated in participants' minds. Second, by focusing sequentially on four steps: (1) providing scientific information on climate change and its effects; (2) identifying response options; (3) prioritizing them; and

(4) drafting an action plan—the debate and potential contentiousness surrounding Steps 3 and 4 is significantly reduced. If the first workshops had begun with a discussion of needed public interventions and policy options to address climate change, the resistance to developing consensus around specific recommendations would likely have been insurmountable (because of grandstanding, the promotion of preconceived agendas, etc.). However, by initially focusing on conveying factual information on climate change and its observed impacts in the region, it proved possible to develop a common understanding among stakeholders regarding the nature of the underlying problems. The involvement of scientists in the first workshops in providing information on climate change and their effects helped reduce the potential for conflicting views. This was because most of the scientists and researchers focused on dispassionate presentations of changes, causes and effects. Most importantly, the participants themselves did not appear to interpret this information as biased.

More Timely and Accessible Meteorological Information Is Needed to Increase Resilience

Jordan and Lebanon should work to improve climate projection information. In the short- and medium-term, the collection and monitoring of climate data could be improved by expanding the number of weather stations, by developing and field-testing different alternatives for remotely sensed indices of drought, and by collaborating with other countries in the region to improve the coverage and comparability of data. This effort should be combined with a push to link climate data with impact analyses by making climate data available to policymakers and researchers. Some efforts in this direction have already begun. For example, Lebanon is part of the European Climate Assessment and Dataset (ECA&D) project. This aims to combine the collation of a daily series of observations at meteorological stations with quality control, an analysis of extremes and the dissemination of both the daily data and the analysis results. This effort to improve climate projection information is gradually being extended across the Middle East.

The accessibility of climate data should be improved in both countries. Several actions can be taken to enhance this accessibility. These include: digitalizing data collected in the past that was stored in formats that can be damaged or difficult to access; and encouraging civil authorities to take responsibility for sharing and making the data available to users. This can be especially important when, for example, meteorological services are under the governance of a Ministry of Defense. Many countries have websites with such data for public use. Still, for security reasons, access to current meteorological data is often limited, but it is important that older data (for example one month or one year) at daily or sub-daily temporal resolution is eventually made public. Ideally, information on the availability, conditions for use, and procedures to access data should be provided and regularly updated (Verner 2012).

Important Policies Can Be Implemented in Both Countries to Build Resilience

Agricultural intensification strategies should be implemented in both countries. There are several central challenges facing the food and agricultural sector in Lebanon, Jordan and elsewhere in the Middle East region. These include food security, rural poverty, the critical role of water-related constraints, urbanization and the resulting loss of farmland, and the vulnerability of rural populations to climate change and price volatility. A common thread to addressing these challenges, not only in the Middle East but elsewhere, is the central importance of successful agricultural intensification strategies (Lee et al. 2001; Vosti and Reardon 1997). Agricultural intensification can be defined as cultivating land to achieve maximum crop output per unit of input. The purpose of these strategies is to increase the productivity and income-generating potential of agriculture on an existing or reduced land base. Numerous investments and interventions are identified in this report to improve agricultural intensification in spite of the challenges presented by climate change. These include: measures to increase the productivity of high-value products, especially fruits and vegetables; improved irrigation and water management; public investments in agricultural research and development; private investments in food marketing and distribution; and a variety of institutional and policy changes to provide a more enabling environment. Each of the activities prioritized in the action plans represent strategies to promote agricultural intensification.

Jordan and Lebanon should focus on improving the production and productivity of "value-added agriculture," particularly that of fruits and vegetables. Value-added agriculture can be generally defined as the processing or manufacturing of an agricultural product to enhance its value. An example would be producing wine from grapes. Such value-added strategies in agriculture meet a number of the criteria critical to development in middle-income countries (Cowan 2003; Meijerink and Roza 2007). These criteria include: a high potential for growth in consumer demand; the proximity of both domestic and export markets; high returns per unit of land (particularly important for small landholders); and high levels of diversification both in terms of production and consumption (by contributing to food security through dietary diversification). In terms of climate change adaptation, particularly in the water-scarce environments of countries like Lebanon and Jordan, value-added agriculture takes on particular importance in terms of its economical use of water inputs, its potential to be successful in areas experiencing urban growth and farmland loss (such as in Lebanon's Bekaa Valley), and its ability to take advantage of the local research base.

No- and low-regret adaptation strategies should be pursued in both Jordan and Lebanon. As is evident in chapter 2 of this report, downscaled climate projections for Lebanon and Jordan demonstrate potentially severe impacts from climate change throughout the twenty-first century. These trends are already underway and are expected to be exacerbated in the future. Yet, despite these projections, it is still highly uncertain how these changes will impact humans. Will there be more floods? If so, where? Will they lead to increased competition over dwindling resources, migration or social conflict? It is because of this uncertainty that it is important to implement strategies that will have net positive social benefits regardless of climate impacts. Such a "no-regrets" approach will assure that maladaptive strategies (adaptation strategies that lead to negative outcomes) are not enacted. In this light, no-regret and low-regret climate adaptation options and policies that generate high direct or indirect benefits currently, even in the face of uncertainty regarding future climate impacts, would be judicious. The action plans from both Jordan and Lebanon include many no- or low-regrets options related to water. These include: improving the irrigation and water delivery infrastructure; research on new water management technologies and crop varieties; increasing water use efficiency through improved management practices; improved climate monitoring and early warning systems; and a variety of institutional changes that better enable farms and rural households to respond to the changing environment. Since water scarcity is a major issue even without climate change, policy makers will have "no regrets" about improving water use efficiency.

Policy makers should consistently consider the input of local stakeholders, and mechanisms should be in place to assure this. Local stakeholders are ultimately those whose livelihoods depend most on the success of strategies and policies related to agricultural adaptation. The impacts of climate change are highly unique to specific localities and, therefore, local people are the most familiar with the on-the-ground realities of their social and agricultural ecosystems (World Bank 2009). Thus, the recommendations and priorities expressed by local stakeholders are particularly important when considering future investments and options to facilitate climate change adaptation. This is reinforced by the fact that the response options identified and prioritized in the action plans echo many of the interventions and strategic investments recommended elsewhere by policy makers, international donors, multilateral organizations and others. These "bottom-up" recommendations help validate and reinforce the strategies made in other contexts, including policy-driven and "top-down" strategies. Strategies to promote the continuing involvement of agricultural stakeholders in moving forward are diverse, but include: (1) giving a role to farmer organizations, watershed councils, and similar local institutions in the promotion and execution of climate strategies; (2) promoting the wider use of on-farm trials (not only at experiment stations under "scientific" conditions) by agricultural researchers, for example, in crop varietal development and the development of integrated pest management (IPM) strategies; and (3) assuring the representation by local farmers and farmer organizations in regional and national agricultural policy formulation. A recent World Bank publication (World Bank 2011a offers a framework to help achieve this).

New technologies should be utilized in the agricultural sector of both countries, with mechanisms in place for continuous revision. Many of the response options prioritized by local stakeholders in this project focus on technological solutions to climate adaptation—improved irrigation technologies, water harvesting and storage, the development of drought-tolerant crop varieties, improving technologies for groundwater and climate monitoring, and so forth. These technological solutions are important, and indeed, some—like the development of drought-tolerant varieties—are often viewed as central to effective climate adaptation in agriculture. A yet-to-be published report by Lebanon's Ministry of Environment (Ministry of Environment (Lebanon), UNEP Risoe Center, and UNDP 2012) specifically prioritizes a number of technologies related to the agricultural sector. These include: conservation agriculture, risk-coping production systems, selection of adapted varieties and rootstocks, integrated pest management, integrated production and protection for greenhouses, early warning systems that incorporate innovative information and communication technologies, and index insurance. Nonetheless, these are not enough (Huesemann 2003). Technological advances are never permanent; they always have a shelf life. Technological change in agriculture is an ongoing process that is key to achieving continuing productivity improvements, whose impacts can be reinforced and magnified through concurrent attention to improving management. Local stakeholders in both countries understand this and, notwithstanding their prioritization of a number of technological interventions and investments, also highlighted the importance of improved management and capacity-building in these technologies. This was indicated through such measures as agricultural extension, dissemination of research results, and the building of human capacity to deal with future climate changes at all levels—on the part of farmers, government officials, researchers, and others.

The public sector should play a major role in climate change adaptation investments, interventions and policy changes. Ultimately, it is the private decision maker and resource manager—primarily among farm households—who must make the key decisions regarding resource allocation. These decisions include what crops to plant, how much to produce, and similar decisions. But the prioritized response options suggest a critical role for the public sector in dealing with climate change adaptation in agriculture. That role has previously been summarized as focusing on the "three I's" (World Bank 2009): Investments, Information, and Institutions and policy innovations. In Lebanon, the response options identified and prioritized by stakeholders include a public-private partnership to develop climate-proof plant materials, the improvement of climate monitoring systems, and establishing a national climate change authority. In Jordan, the final priority list of response options includes a number of proposed public policy options—a national climate change strategy, developing land-use laws, and the implementation of an agricultural insurance scheme—that could fundamentally affect the overall environment and incentive structure for private decision making over resource use in agriculture.

The public sector has a role in improving the information base available to farmers and farm households. The World Bank's Middle East and North Africa Region's Flagship report on climate change (World Bank 2012a) describes how there is a lack of quality information and data collection activities related to the climate in the Arab region. Even when data are collected they are not consolidated or are unavailable. Yet, individual farmers and rural households would benefit greatly from the use of this information. This continues to be true in the context of climate change adaptation. Much of the information base on which farmers make their private resource decisions is not available but can be considered as a public good—non-excludable and non-rivalrous in demand (Cook and Sachs 1999). As a public good, there is commonly an under-supply of information by the private sector. As a result, sub-optimal resource allocation and management decisions are common with negative impacts on production, productivity and food security. For these basic reasons, many institutions have prioritized information provision and decision support as a key mechanisms for public sector investments in adaptation (World Bank 2009). Stakeholders in Lebanon and Jordan have prioritized response options that would be strengthened by public sector support. These include, in Lebanon: climate monitoring systems and databases, and technical advice on irrigation management and integrated pest management; and in Jordan: improvements in the early warning system for drought, and improvement of the information base on crop management practices and water resource management. The public sector can potentially play a key role in supplying this type of information, thus improving the capacity of farmers and resource managers to address the challenges posed by climate change.

Road Map of This Report

This report is comprised of four chapters and is written to be accessible to a general audience, but with sufficient technical detail to be of interest to specialists and experts alike in the fields of climate science and climate change adaptation. The first chapter is the introduction. The second chapter, which will be of particular interest to climate scientists, is a technical exercise in downscaling climate data for Jordan and Lebanon that then uses this information to predict future trends. This provides climate risk information (from both historical data and climate scenarios for the region at high time and space resolution for the 2020s, 2050s and 2080s). In due course, it is anticipated that these climate scenarios will feed into water management and agro-economic models to help assess potential impacts of climate change on the water and agricultural sectors. The third chapter, which will be of interest to practitioners and policy makers in Jordan and Lebanon, summarizes the results of the priority-setting methodology employed in this study to identify stakeholder-based strategies to build agricultural resilience in selected regions of both countries. This information is being used in an ongoing dialogue with officials in both Lebanon and Jordan. The chapter has separate sections for Lebanon and Jordan, so individuals interested in one country or the other can easily read sections pertaining to each. The fourth and final chapter is a summary of conclusions as well as a number of practical recommendations for policy makers stemming from the rest of the report.

References

Agrawal, A., C. McSweeney, and N. Perrin. 2008. "Local Institutions and Climate Change Adaptation." The Social Dimensions of Climate Change No. 113, World Bank, Washington, DC, July.

Cook, L. D., and J. Sachs. 1999. "Regional Public Goods in International Assistance." In *Global Public Goods: International Cooperation in the 21st Century*, edited by I. Kaul, I. Grunberg, and I. Stern. Oxford; New York: Oxford University Press for the United Nations Development Programme.

Cowan, T. 2003. *Value-Added Agricultural Enterprises in Rural Development Strategies.* Hauppauge, New York: Nova Science Pub., Inc.

FAO (Food and Agriculture Organization). 2011. "Regional Priority Framework for the Near East." FAO Regional Office for the Near East, Rome, Italy. http://neareast.fao.org/FCKupload/File/RPF-EN.pdf.

Huesemann, M. H. 2003. "The Limits of Technological Solutions to Sustainable Development." *Clean Technologies and Environmental Policy* 5: 21–34.

IPCC (Intergovernmental Panel on Climate Change). 2007a. *Impacts, Adaptation and Vulnerability, Contribution of Working Group II to Fourth Assessment Report of the Intergovernmental Panel on Climate Change.* Cambridge, UK: Cambridge University Press.

———. 2007b. "Special Report on the Regional Impacts of Climate Change: An Assessment of Vulnerability." Geneva, Switzerland.

IMF (International Monetary Fund). 2010. "Lebanon: Real GDP Growth Analysis." Resident Representative Office in Lebanon, July. http://www.imf.org/external/country/LBN/rr/2010/070110.pdf.

Lampietti, J. 2010. "The Future of Agriculture in Lebanon under Climate Change." Development Horizons, Middle East Department, World Bank, First/Second Quarter, 12–14.

Lee, D. R., C. B. Barrett, P. Hazell, and D. Southgate. 2001. "Assessing Tradeoffs and Synergies among Agricultural Intensification, Economic Development and Environmental Goals: Conclusions and Implications for Policy." In *Tradeoffs or Synergies? Agricultural Intensification, Economic Development and the Environment*, Ch. 24, edited by D. R. Lee and C. B. Barrett. Wallingford, UK: CABI Publishing.

Meijerink, G., and P. Roza. 2007. "The Role of Agriculture in Economic Development." Markets, Chains and Sustainable Development Strategy & Policy Paper 4, Stichting DLO, Wageningen.

MOA (Ministry of Agriculture [Lebanon]). 2004. "Strategie du Developpement Agricole au Liban." République Libanaise Ministère de l'Agriculture. Direction des Etudes et de la Coordination. http://www.loadleb.org/files/strategy/strategy/1_STRATEGIE%20AGRICOLE%20FINAL_Fr.pdf.

Ministry of Environment (Lebanon), UNEP Risoe center, and UNDP. 2012. "Technology Needs Assessment." Ministry of Environment, Beirut, Lebanon. Unpublished report.

Mitchell, T., and S. Maxwell. 2010. "Defining Climate Compatible Development." Policy Brief, Climate & Development Knowledge Network, London, November. http://cdkn. preprod.headshift.com/wp-content/uploads/2011/02/CDKN-CCD-DIGI-MASTER-19NOV.pdf.

Nelson, G. C., M. W. Rosegrant, J. Koo, R. Robertson, T. Sulser, T. Zhu, C. Ringler, S. Msangi, A. Palazzo, M. Batka, M. Magalhaes, R. Valmonte-Santos, M. Ewing, and D. Lee. 2009. *Climate Change: Impact on Agriculture and Costs of Adaptation.* Washington, DC: IFPRI Food Policy Report, International Food Policy Research Institute.

Padgham, J. 2009. "Agricultural Development Under a Changing Climate: Opportunities and Challenges for Adaptation." Joint Discussion Paper, Agriculture and Rural Development & Environment Departments, World Bank, Washington, DC. Verner, D. 2012. Adaptation to a Changing Climate in the Arab Countries.

Verner, D., ed. 2012. "Adaptation to a Changing Climate in the Arab Countries: A Case for Adaptation Governance and Leadership in Building Climate Resilience." MENA Development Report, World Bank Publications, Washington, DC.

Vosti, S. A., and T. Reardon, eds. 1997. *Sustainability, Growth and Poverty Alleviation: A Policy and Agroecological Perspective.* Baltimore: Johns Hopkins University Press.

World Bank. 2007. *World Development Report 2008.* Washington, DC: Agriculture for Development.

———. 2008. *World Development Report 2008.* Agriculture for Development. Washington, DC: World Bank. http://siteresources.worldbank.org/INTWDR2008/Resources/WDR_00_book.pdf.

———. 2009. *Building Response Strategies to Climate Change in Agricultural Systems in Latin America.* Washington, DC: Latin American and the Caribbean Region.

———. 2011a. "The Adaptation Coalition Framework: Building Community Resilience to Climate Change." Social Development Unit of Latin America and Caribbean Region, World Bank, Washington, DC. http://siteresources.worldbank.org/EXTSOCIALDEVELOPMENT/Resources/244362-1232059926563/5747581-1239131985528/Adaptation-Coalition-Toolkit_Building-Community-Resilience-Climate-Change_web.pdf.

———. 2011b. "Mainstreaming Adaptation to Climate Change in Agriculture and Natural Resources Management Projects." World Bank, Washington, DC. http://climatechange.worldbank.org/content/mainstreaming-adaptation-climate-change-agriculture-and-natural-resources-management-project revised entry.

———. 2012a. "Adaptation to Climate Change in the Middle East and North Africa Region." MENA Region, World Bank, Washington, DC. http://web.worldbank.org/WBSITE/EXTERNAL/COUNTRIES/MENAEXT/0,contentMDK%3A21596766~pagePK%3A146736.

CHAPTER 1

Introduction

Photograph by Dorte Verner

An increasing sense of urgency characterizes the global dialogue on climate change. The landmark Fourth Assessment Report of the Intergovernmental Panel on Climate Change (IPCC 2007a) concludes definitively that "[w]arming of the climate system is unequivocal" and goes on to elaborate the expected pervasive effects of future climatic changes—rising air and ocean temperatures, extensive melting of snow and ice, rising sea levels, and regional impacts ranging from modest to severe. The World Bank's *World Development Report 2010* (World Bank 2009b) argues that rising greenhouse gases have fundamentally "transformed the relationship between people and the environment," and that, "left unmanaged, climate change will reverse development progress and compromise the well-being of current and future generations." The implications for people in low-income countries are expected to be particularly severe, given their greater exposure to natural hazards, disasters and sources of environmental risk, their often already precarious livelihoods, and a frequent absence of economic and institutional buffers to moderate the impacts of future climate change.

The climate challenges confronting development in the Middle East are particularly stark. This region, and in particular its rural people, face what might be called a "triple threat" from climate change. First, the Middle East is already one of the driest and most water-scarce regions of the world (World Bank 2011a) and faces severe challenges posed by high temperatures and limited water supplies. Higher temperatures, even with no changes in precipitation, can be expected to increase soil moisture evaporation and thus pose the risk of increased soil degradation and other effects (IPCC 2007b). Second, although IPCC (2007b) projections of future changes in precipitation trends in the Eastern Mediterranean region are uncertain, there is a "strong consensus" that annual precipitation totals will fall, and it is estimated that precipitation could be reduced by 10–20 percent (Wilby 2010). Such declines would further exacerbate the already precarious water scarcity facing the inhabitants of much of the region, who face severe water constraints for both human and agricultural use. Third, the rural economies of the region, as well as many rural people, are highly dependent on agriculture, a sector that is itself extremely vulnerable to climate change. The sources of agricultural vulnerability are several-fold, including reduced yields, the likelihood of crop failures due to extreme weather events (droughts, floods, extreme temperatures), increasing incidence of crop pests and disease, and so on. In the Middle East, a recent study by the International Food Policy Research Institute suggests that climate change impacts on crop yields (to the year 2050) will be particularly severe with reductions of 30 percent for rice, 47 percent for maize, and 20 percent for wheat (Nelson et al. 2009).

For all these reasons—the dependence of their livelihoods on climate and natural resources, their vulnerability to rising temperatures and water scarcity, and the increased likelihood of future droughts and severe weather events—farmers and rural households are already on the "front lines" of dealing with and adapting to climate change. Farmers constantly make autonomous adaptations in the face of ongoing changes to agroecosystems, not just climate change. These autonomous adaptation strategies include changing crop selection, allocating land between alternative crop and livestock uses, and adjusting planting and harvesting dates. The rural poor are especially vulnerable to climate change impacts due to their disproportionate dependence on agriculture, lower ability to adapt, and frequent lack of support systems to facilitate adaptation (World Bank 2008). Grasslands, livestock, and water resources in the region—and the rural households dependent on these resources—are expected to be particularly vulnerable to climate change (IPCC 2007a). Beyond the level of individual farms and households, the impacts of climate change on local social structures and agroecosystems in the Middle East and elsewhere will be significant. Land degradation may result from decreased soil moisture and overgrazing. In addition to the impacts on productive capacity, climate change may also trigger major effects on health and food security in the region. Heat stress and increases in vectorborne and waterborne disease are expected to occur. Other impacts include decreases in caloric availability of up to 500 calories per person per day and in an increase of one million malnourished children in the North Africa and Middle East compared to a "no climate change" scenario (Nelson et al. 2009).

To deal with these growing constraints, effective adaptation is critical. In addition to autonomous adaptations, agricultural climate adaptation can be planned (FAO 2007). Planned adaptations, however, are deliberate long-term response strategies and policy interventions developed in response to the recognition of changed conditions and aimed at improving the resilience of agroecosystems (FAO 2007). These adaptations can include the spontaneous adaptations mentioned above but can also go well beyond these to encompass research and technology development, investments in irrigation systems, genetic improvements and crop varietal development, and institutional changes in intellectual property rights, land-use and agricultural policy. Because many of the impacts of climate change on agriculture are predictable—qualitatively if not quantitatively—planned adaptation is widely considered to be a prudent and efficient response, and many organizations have urged that financial resources be devoted to investing in agricultural adaptations (Nelson et al. 2009; World Bank 2007). The public sector has a significant role to play in making these investments and providing an institutional and policy framework that facilitates adaptation strategies. Since many of these investments are part of good agricultural development policy anyway, they can yield "double dividends," in that they not only represent valuable and appropriate measures to foster agricultural development but they also contribute to agroecosystem resilience to sources of climate risk (Padgham 2009). Ultimately, the effectiveness of adaptation strategies depends on how well they deal with the diverse local conditions faced by farm households and rural communities (Agrawal, McSweeney, and Perrin 2008; World Bank 2009a).

The challenge in developing agricultural adaptation strategies to locally relevant solutions frequently entails addressing two interrelated goals. These are, first, decreasing the vulnerability of local communities to climate variability and climate change, and second, increasing the resilience of natural and social systems in the face of climate change (World Bank 2011b). Developing solutions to address these goals, however, quickly encounters two major challenges. As agriculture is a natural resource–based industry and one heavily dependent on varying local conditions in climate, soils, and other physical and community characteristics, the first challenge in agricultural adaptation is to identify specific, locally appropriate solutions to climatic change that address specific local needs. In addition, since investing in adaptation responses invariably entails costs and resources are inevitably constrained, the second challenge is one of prioritizing the options that are available in a manner that effectively addresses climate change while making the best use of resources and available options.

This report is comprised of four chapters including this Introduction. The second chapter is a technical exercise in downscaling climate data for Jordan and Lebanon and then using this information to predict future trends. This provides climate risk information (from both historical data and climate scenarios for the region at high time and space resolution for the 2020s, 2050s, and 2080s). In due course, it is anticipated that these climate scenarios will feed into water management and agro-economic models to help assess potential impacts of climate change on the water and agricultural sectors. The third chapter is a presentation of the results from priority-setting strategies to build agricultural resilience implemented in select regions of both countries. This information is being used in an ongoing dialogue with officials in both Lebanon and Jordan. The fourth and final chapter is a summary of conclusions as well as a number of practical recommendations for policy makers stemming from the rest of the report.

This report aims to assist the people of Jordan and Lebanon in understanding the specific challenges and opportunities posed by climate change in the agricultural sector and to develop local-level priorities, informed by stakeholder input, to build agricultural resilience in Jordan (Jordan River Valley) and Lebanon (Bekaa Valley). The objectives of this study were threefold: (1) to improve the understanding of climate change projections and impacts on rural community livelihoods in select regions of Jordan and Lebanon; (2) to engage local communities, farmers, local experts, and local and national government representatives in a participatory fashion in helping craft adaptation agricultural response options to climate change; and (3) to develop local and regional climate change action plans that formulate recommendations for investment strategies and interventions in local agricultural systems. This report may serve as the analytical underpinning for ongoing discussions taking place within each ministry on how to best move forward in building agricultural resilience to climate change.

In order to develop the local-level priorities for agricultural adaptation to climate change, a four-step priority-setting process was used. This process was built around a series of workshops and meetings with stakeholders and policy makers (described in detail in the "Methodology" section in chapter 3). Two country teams led the activities in each country, including the preparation of the Action Plans. In Jordan, the country team was led by the National Center for Agricultural Research and Extension (NCARE). In Lebanon, the Lebanese Agricultural Research Institute (LARI) undertook the central organizing role for the study. Scientists, researchers, and officials from national government institutions, universities, and other organizations played active roles as presenters and advisers at the workshops. Others who participated in these workshops included farmers and representatives of farmer organizations, extension specialists, representatives from NGOs and local governments, journalists, and representatives from international development and research organizations. World Bank staff and consultants also played an important supporting role at the workshops.

It is important to take note of several aspects of this study. First, the focus is on climate adaptation, not mitigation. As mentioned below, the two countries that are included in this study, Jordan and Lebanon, are minimal contributors to total global greenhouse gas (GHG) emissions. While efforts are under way in each country to curtail emissions, and these efforts should be supported and expanded, the primary challenges in the agricultural sector with respect to climate change faced in both countries are ones of adaptation. As suggested in other contexts, many "climate compatible" adaptation options may simultaneously reduce GHG emissions and promote mitigation (Mitchell and Maxwell 2010). However, the main thrust of this project is identifying and prioritizing local agricultural adaptation strategies and response options in the two regions.

Second, the approach followed in this study emphasizes the participation of diverse local stakeholders in formulating response options that address needs in climate change adaptation in agriculture. This "bottom-up" approach is an important complement to the frequent "top-down" approaches taken in many climate change strategies and national action plans. To some extent such top-down approaches are inevitable in dealing with mitigation strategies, which necessarily involve tradeoffs among sectors and regions, and where establishing national goals and policies to achieve those goals is required. Yet, in formulating adaptation strategies, the local focus is indispensable, so it is essential that any methodology or approach must involve the input of local stakeholders in a bottom-up fashion.

Third, it should be emphasized that the agricultural adaptation response options identified and prioritized in this study, and the local and regional action plans that build on those response options are only a first step in generating the needed investments, policy changes and changes in private sector decision making. As in previous applications of the priority-setting methodology employed in this study (see World Bank, 2009a), the overall objective of this project was to identify priority interventions, based on local stakeholder inputs, that will facilitate building climate resilience in agriculture. The objective was not to conduct a comprehensive economic feasibility analysis or cost-benefit analysis of the response options considered. The priority-setting approach is not a substitute for rigorous economic assessment of possible investments and interventions; rather, the two approaches are complementary. Further detailed technical and economic analyses are a necessary second step, particularly in cases where major public investments are involved, such as in the construction of new irrigation infrastructure or establishing climate early warning systems.

As a result of the climate downscaling and modeling (chapter 2 of this report), several conclusions were made for both countries. These include the following:

- There have been rising air temperatures since the 1970s, at a rate of 1.8–3.6°C per century.
- There have been reductions in annual rainfall, by –12 percent in Lebanon since the 1980s, and –5 percent to –20 percent in Jordan since the 1960s.
- The frequency of hot nights has risen.
- Heavy-rainfall events have become more frequent and intense.

- Average river flows and groundwater levels have generally declined.
- The Arab region, including Lebanon and Jordan, will warm more rapidly than the global average, by a projected ~1.3°C–2.7°C by mid-century, and annual precipitation totals will fall as much as 10–20 percent.
- The imbalances created by shrinking water supplies and rising demands will be exacerbated.
- These trends are predicted to continue over the remainder of this century.

In chapter 3, which describes the process and results of the priority-setting methodology, stakeholders in each study region ranked a number of activities that should be undertaken to increase agricultural resilience to climate change. These are summarized in order of highest-to-lowest priority in table 1.1 below.

In the final chapter (chapter 4), it is concluded that the priority-setting methodology was successful in achieving its aims by creating a space for local stakeholders to participate in adaptation decision making. Some of the limitations of this approach are also discussed. Based on these conclusions and the preceding two chapters, several recommendations are made. These include the following:

- Agricultural intensification strategies should be implemented in both countries.
- Jordan and Lebanon should focus on improving the production and productivity of "value-added agriculture," particularly that of fruits and vegetables.
- No- and low-regret adaptation strategies should be pursued in both Jordan and Lebanon.
- Policy makers should consistently consider the input of local stakeholders, and mechanisms should be in place to assure this.

Table 1.1 Summary of Action Plans for Lebanon and Jordan in 2012

Lebanon	Jordan
1. Adopt new irrigation technologies	1. Improve farm production systems and productivity
2. Launchproject to construct small- and medium-scale water harvesting reservoirs	2. Improve on-farm water use efficiency and integrated water resources management
3. Integrate the production management of pests, diseases and plant disorders under climate change	3. Improve livestock and rangeland systems
4. Produce and distribute plant materials adapted to climate change	4. Build national capacity for climate change adaptation
5. Increase capacity for climate change adaptation	5. Reduce risks of agricultural pests and diseases
6. Evaluate and maintain the genetic diversity of wild species and local varieties adapted to climate change	6. Reinforce early warning system for drought
	7. Reform land-use laws and implement sustainable land-use
	8. Activate an agricultural risk management fund

Source: World Bank data.

- New technologies should be utilized in the agricultural sector of both countries with mechanisms in place for continuous revision and to utilize new advancements.
- The public sector should play a major role in climate change adaptation investments, interventions, and policy changes.
- The public sector has a role in improving the information base available to farmers and farm households.

References

Agrawal, A., C. McSweeney, and N. Perrin. 2008. "Local Institutions and Climate Change Adaptation." The Social Dimensions of Climate Change No. 113, World Bank, Washington, DC, July.

FAO (Food and Agriculture Organization). 2007. *Adaptation to Climate Change in Agriculture, Forestry and Fisheries: Perspective, Framework and Priorities*. Rome, Italy: Interdepartmental Working Group on Climate Change.

IPCC (Intergovernmental Panel on Climate Change). 2007a. *Impacts, Adaptation and Vulnerability, Contribution of Working Group II to Fourth Assessment Report of the Intergovernmental Panel on Climate Change*. Cambridge, UK: Cambridge University Press.

IPCC (Intergovernmental Panel on Climate Change). 2007b. "Special Report on the Regional Impacts of Climate Change: An Assessment of Vulnerability." Geneva, Switzerland.

Mitchell, T., and S. Maxwell. 2010. "Defining Climate Compatible Development." Policy Brief. Climate & Development Knowledge Network, London, November. http://cdkn.preprod.headshift.com/wp-content/uploads/2011/02/CDKN-CCD-DIGI-MASTER-19NOV.pdf.

Nelson, G. C., M. W. Rosegrant, J. Koo, R. Robertson, T. Sulser, T. Zhu, C. Ringler, S. Msangi, A. Palazzo, M. Batka, M. Magalhaes, R. Valmonte-Santos, M. Ewing, and D. Lee. 2009. "Climate Change: Impact on Agriculture and Costs of Adaptation." IFPRI Food Policy Report, International Food Policy Research Institute, Washington, DC.

Padgham, J. 2009. "Agricultural Development Under a Changing Climate: Opportunities and Challenges for Adaptation." Joint Discussion Paper, Agriculture and Rural Development & Environment Departments, World Bank, Washington, DC.

Wilby, R. L. 2010. "Climate Change Projections and Downscaling for Jordan, Lebanon and Syria: Draft Synthesis Report." World Bank, MENA Region, September.

World Bank. 2007. *World Development Report 2008: Agriculture for Development*. Washington, DC: World Bank.

———. 2008. *World Development Report 2008: Agriculture for Development*. Washington, DC: World Bank. http://siteresources.worldbank.org/INTWDR2008/Resources/WDR_00_book.pdf.

———. 2009a. *Building Response Strategies to Climate Change in Agricultural Systems in Latin America*. Latin American and the Caribbean Region. Washington, DC: World Bank.

———. 2009b. *World Development Report 2010: Development and Climate Change*. Washington, DC: World Bank.

———. 2011a. *The Adaptation Coalition Framework: Building Community Resilience to Climate Change.* Social Development Unit of Latin America and Caribbean Region. Washington, DC: World Bank. http://siteresources.worldbank.org/ EXTSOCIALDEVELOPMENT/Resources/244362-12320599 26563/5747581-1239131985528/Adaptation-Coalition-Toolkit_Building-Community-Resilience-Climate-Change_web.pdf.

———. 2011b. "Mainstreaming Adaptation to Climate Change in Agriculture and Natural Resources Management Projects." World Bank, Washington, DC. http://climatechange. worldbank.org/climatechange/content/mainstreaming-adaptation-climate-change-agriculture-and-natural-resources-management-project revised entry.

CHAPTER 2

Climate Change in Lebanon and Jordan

Photograph by Dorte Verner

This chapter provides a synthesis of evidence of climate variability and change in Jordan and Lebanon. The work is intended to support strategic planning for agriculture, irrigation, and rural development within the region. More specifically, climate projections like those reported here are used to backstop the participatory stakeholder-based response options that are identified and discussed in subsequent chapters of this report. In this manner, climate modeling provides a science-based mechanism to reinforce and supplement—and occasionally, to offset—local knowledge and stakeholder experience in helping guide short- and long-term investments and interventions to address climate adaptation.

The evaluation summarized here is based on a range of primary and secondary data sources held by national and international agencies, including conventional meteorological records, derived climate indices, and satellite products. This data supports the view that the region has experienced rising air temperatures since the 1970s remove boldface throughout this paragraph and the next, and shrinking snow cover. There have also been local reductions in rainfall since the 1960s, set against a background of large inter-annual variability linked to the behavior of the North Atlantic Oscillation. Trends in extreme events are more problematic to assess, but there are indications that the frequency of hot nights has risen, and that heavy rainfall events have become more frequent and intense. Average river flows and groundwater levels have generally declined but it is difficult to disentangle climate-driven trends from those caused by growth in water abstractions over the same period.

Regional climate change projections suggest that these trends are set to continue over the coming decades, potentially exacerbating the imbalance between water supply and demand. There is strong consensus amongst climate models that the region will warm more rapidly than the global average and that annual precipitation totals will fall. By the 2050s mean temperatures could increase by ~1.5°C and precipitation could be reduced by 10–20 percent. Future changes in rainfall will reflect the interplay between a possible northward shift in the Mediterranean storm track, counteracted by rising sea surface temperatures and more frequent polar intrusions. Weaker and/or less frequent west-east airflows would tend to reduce orographic rainfall in the region's mountain ranges.

A statistical downscaling model was used to evaluate site-specific outcomes of the changes in atmospheric conditions. The model was calibrated and tested for 10 locations representing diverse physiographic conditions and agro-ecological zones. Daily mean temperatures and precipitation totals were downscaled from the UK Met Office HadCM3 climate model under two emissions scenarios (SRES A2 and B2) for the period 1961–2099. Downscaled changes in seasonal mean temperatures and precipitation amounts bracket the findings of earlier studies.

Overall, there is a tendency for more rapid warming at higher elevations and with distance from the coast. Warming is most pronounced in spring at coastal sites and for summer at locations inland. The largest reductions to annual rainfall are found for sites in the coastal zone, and within the Bekaa Valley, where changes could be in the range 10–30 percent by the 2050s and 20–50 percent by the 2080s. Discerning changes in precipitation for the 2020s is generally problematic due to climate model and downscaling biases, combined with large natural variability in annual totals.

A major strength of statistical downscaling is the ability to derive indices of change that are meaningful to planners such as the likelihood of dry-years, winter growing degree days, or dry season duration. A case study for Amman (Jordan) showed that the chance of a dry year (<200 millimeters) was historically one in three years, but could increase to one in two years (that is a 50 percent chance) by the late 2020s. Extension of the dry-season duration by ~30 days by the 2050s could limit the length of grazing time. The focus of this chapter is on changes in temperature and precipitation over the study sites.

However, in addition sea level rise could increase the risk of saline intrusion to coastal aquifers further limiting the resources available to irrigators and urban areas such as Beirut. More intense rainfall could also limit the effectiveness of groundwater recharge. Preliminary research undertaken elsewhere indicates that the future water security of the region will hinge, in part, on changes in snowpack and residual river-flows in the Euphrates, and on negotiated water-sharing with Israel. Likewise, future food security will depend on the ability of domestic and global producers to adapt to changing conditions. Therefore, any sector planning will need to take account of potential impacts of climate change on the environment, water politics and agro-economics of the wider regional neighborhood.

The regional climate change projections and uncertainties described below are a cause for concern given existing water supply deficits and a legacy of over-exploited freshwater stocks. However, in the short and medium term, population and economic growth are more important drivers of the water deficit than climate change. Possible exceptions include situations where a tipping point (such as the limit to rain-fed agriculture ~200 millimeters per year) is being approached. Under more rapid climate change the threshold would be reached sooner, so uncertainty in the climate modeling can translate into uncertainty about the time-scale available for anticipatory adaptation. Further research is underway to better characterize the uncertainty in regional projections by downscaling from more climate models and emissions scenarios.

Global Climate Change

The scientific consensus voiced through Working Group I of the Fourth Assessment Report (FAR) of the Intergovernmental Panel on Climate Change (IPCC) (Solomon et al. 2007) asserts that warming of the climate system is unequivocal. Furthermore, most of the observed increase in globally averaged temperatures since the mid-twentieth century is very likely due to observed increases in anthropogenic greenhouse gas concentrations. Indeed, human influences are now detectable at continental scales in some temperature and wind records.

Even if emissions of greenhouse gases were kept at year 2000 levels, the slow response of the oceans to past emissions will result in further warming of ~0.1°C per decade over the next two decades. Climate model projections indicate that decadal-average warming by 2030 is insensitive to the choice of emission pathway and is very likely to exceed natural variability observed during the twentieth century. Rates of twenty-first century warming over the eastern Mediterranean and Middle East are expected to be greater than the global average.

Although there is high confidence in projected patterns of warming, and that sea levels will continue to rise, there is less certainty about regional rainfall. In general, the amount of precipitation is projected to increase at high latitudes and decrease over most subtropical land regions (map 2.1). However, there is a lack of both reliable baseline data and climate model consensus for large areas

**Map 2.1 Projected Patterns of Precipitation Changes (%) for 2090–99 Compared with 1980–90
Based on Multi-Model Average Projections under the SRES A1B Scenario**

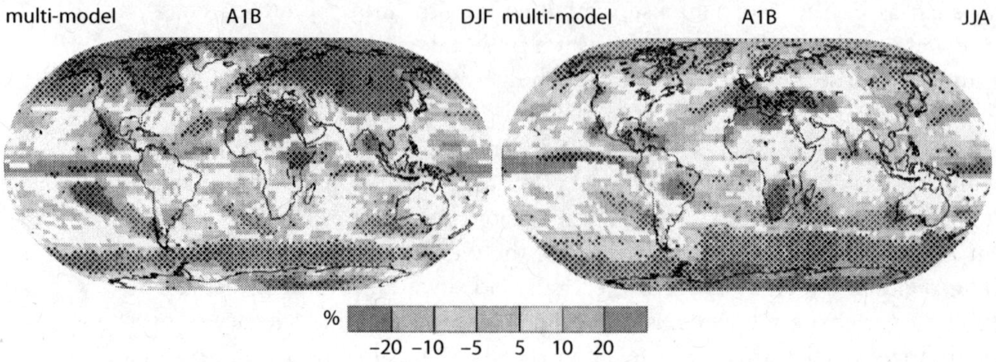

Source: Solomon et al. 2007, Figure TS. 30, p. 76.
Note: White areas show where the model consensus about the sign of the change is less than 66 percent; stippled areas where 90 percent of models agree about the sign. DJF refers to December, January, February, JJA refers to June, July, August.

of Africa, Asia, the Middle East, and South America. The Mediterranean basin is one region where climate model consensus is relatively strong, with the majority of models showing on average decreased winter, summer, and annual precipitation totals (map 2.1).

Climate Information Sources

The Middle East is a relatively data-sparse region. The World Meteorological Organisation (WMO) lists just 11 active meteorological stations for Jordan and 7 for Lebanon.[1] Across the Eastern Mediterranean as a whole there are ~170 stations with monthly temperature and rainfall records (map 2.2, left panel) and in places network densities may be less than one station per 1° × 1° latitude-longitude cell (map 2.2, right panel).

Climate information for the region was compiled from several sources because of the limited availability of raw data. These sources include:

- *Daily weather archives* held by national ministries, meteorological agencies, and agriculture research stations; subject to formal agreement and quality assurance.
- *Secondary data sources and syntheses* of regional case studies as in Zereini and Hötzl (2008). For example, Shahin (2007) provides a compendium of meteorological, hydrological and water quality data for stations across the Arab region.
- *Historical station records* held in global archives. For example, the KNMI Climate Explorer and the NOAA-NCDC Global Surface Summary of Day[2] provide access to daily temperature and rainfall records in the region, Had-CRUT3 gives monthly mean temperature series used in calculations of global

Map 2.2 Location of Rainfall Stations (Left Panel) with Data for 1961–90 and the Average Number of Rain Gauges per 1.0° × 1.0° Grid during 1998–2007 (Right Panel).

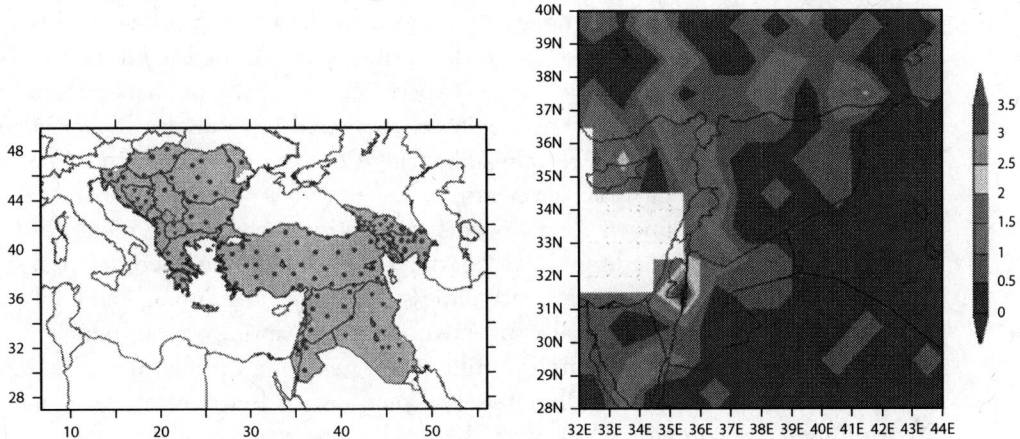

Source: The images and data used in this study were acquired using the GES-DISC Interactive Online Visualization and analysis Infrastructure (Giovanni) as part of the NASA's Goddard Earth Sciences (GES) Data and Information Services Center (DISC)

mean temperature, and the Global Historical Climatology Network Version 2-NCDC (GHCN2) holds monthly rainfall totals (see map 2.2, left panel).

- *Merged satellite-gauge products* such as the Global Precipitation Climatology Project (GPCP). Also, the Tropical Rainfall Measuring Mission (TRMM) multi-satellite precipitation analysis has provided 3-hourly, 0.25° × 0.25° latitude-longitude resolution data in near real time since 1998 (Huffman et al. 2007).

- *Research publications* such as Zhang et al. (2005) and Sensoy et al. (2007) provide information on trends for derived indices of climate extremes for sites across the Middle East. Observed climate indices are available for a few stations in the region via the European Climate Assessment and Dataset[3] (Haylock et al. 2008).

- *Monthly mean gridded climate variables* such as the Climatic Research Unit (CRU) TS 3.0[4] archive provides global coverage at 0.5° × 0.5° latitude-longitude resolution for the period 1901–2006.

- *Climate model output* from the IPCC FAR experiments may be acquired via the Climate and Environmental Retrieving and Archiving (CERA) portal.[5] Downscaling predictor variables are accessed from the Canadian Climate Change Scenarios Network (CCCSN)[6] for a limited number of models. Re-analysis data (quasi-observational airflow and humidity indices) may be obtained from the same source for grid-boxes covering the region (see map 2.5 below).

- *Climate summaries* such as the country profiles of Mitchell, Hulme, and New (2002) provide long-term climate averages and seasonal means for the climate model scenarios (in this case, used in the IPCC Third Assessment Report). Önol and Semazzi (2009) also provide country-average seasonal temperature and precipitation statistics for the present climate (1961–90) based on observations and Regional Climate Model (RCM) simulations.

Climate of the Region

The climate of the region is generally defined as humid "Mediterranean" with cool, wet winters and hot, dry summers, modulated by distance from coast and elevation (Goldreich 1994). Sites close to the eastern shore of the Mediterranean Sea (for example, Beirut, Lattakia, Tripoli) experience a narrower annual temperature range (~15°C) and relatively mild winters compared to locations further inland (for example, Deir ez Zor, El Qamishli) where the annual temperature range is ~25°C. Maximum daytime temperatures exceeding 50°C have been recorded near the Dead Sea. Snowfall only occurs at elevations above 1,000 meters.

Away from the Mediterranean coastal zone the climate becomes progressively more arid with very hot and rainless conditions in summer, and cool wet winters. There are occasional convective storms in spring during which peak rainfall intensities may reach 100 millimeters per hour. Locally, annual rainfall totals can exceed 1,400 millimeters per year east of Betroun in the Lebanon Mountains, declining to less than 300 millimeters per year 150 kilometers further east (map 2.3). The desert zones of eastern Jordan can receive less than 100 millimeters per year.

Map 2.3 Accumulated Rainfall Totals (Millimeters) Based on Monthly GPCP for the Decade 1998–2007

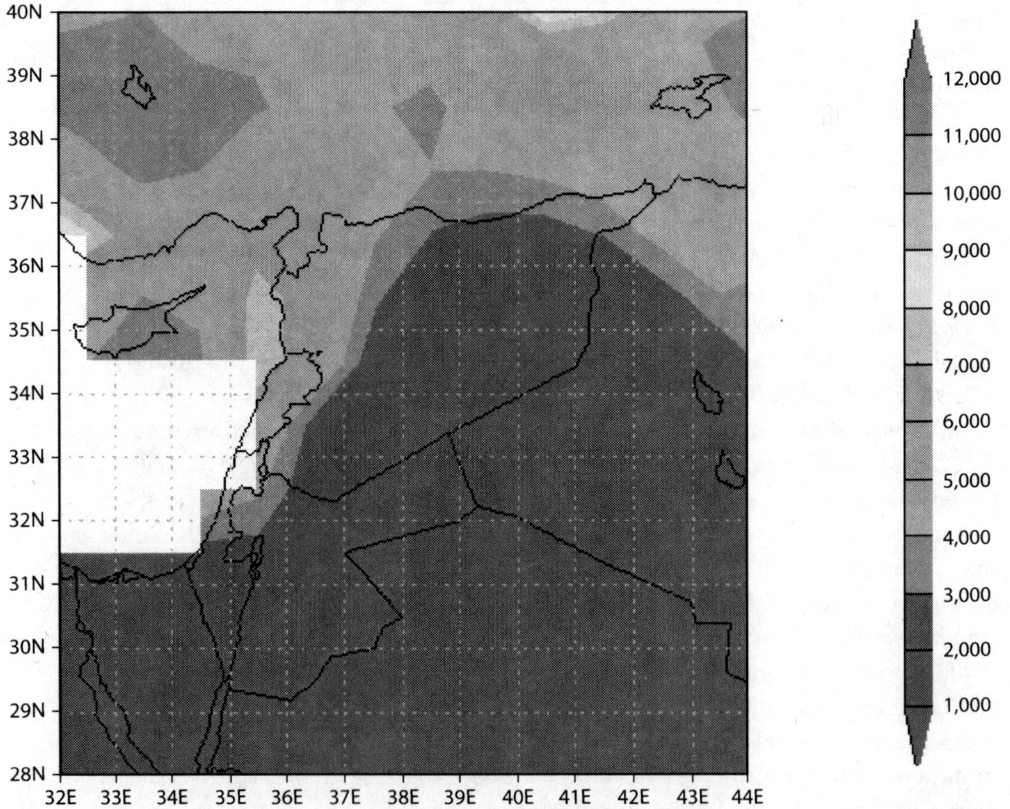

Source: The images and data used in this study were acquired using the GES-DISC Interactive Online Visualization and analysis Infrastructure (Giovanni) as part of the NASA's Goddard Earth Sciences (GES) Data and Information Services Center (DISC)

Strong spatial gradients in rainfall are clearly evident across the region (map 2.3) reflecting the orographic and coastal influences noted above. The interior deserts of Jordan would be expected to be dry because of their latitude, but the surrounding Taurus and Zagros mountain ranges, which feed the Euphrates and Tigris Rivers, amplify this effect (Evans, Smith, and Oglesby 2004). Inter-annual variations in the North Atlantic Oscillation (NAO) are also known to influence the dominant storm-track across the region and hence the frequency of winter cyclones—the dominant source of rainfall in the Middle East (MedCLIVAR 2004). Negative phases of the NAO are associated with a belt of above average rainfall that is most pronounced around latitude 40°N (Cullen and deMenocol 2000; Cullen et al. 2002; Eshel and Farrell 2000; Xoplaki et al. 2004; Ziv et al. 2006).

TRMM[7] data provides a valuable resource for exploring the detail of space-time variations in rainfall across the region. For example, map 2.4 compares winter rainfall totals under positive (+0.65) and strongly negative (−1.67) NAO indices. As expected, the former has lower accumulations with most of Jordan and Lebanon receiving less than 300 millimeters.

As well as providing information on rainfall behavior over annual and decadal time-scales, TRMM data can also be compared with point measurements to assess the character of short-duration events and the feasibility of rainfall estimation at sites not covered by meteorological networks. Two sites are used to demonstrate the potential of the technology: Kamishli (Syria) and Amman (Jordan) (shown as white dots in map 2.5).

Figure 2.1 shows the TRMM estimate of daily rainfall totals for the cell closest to Kamishli (pixel 37–37.25°N, 41–41.25°E) and Amman (pixel 31.75–32°N, 35.75–36°E). At both sites, TRMM yields realistic rainfall amounts, clustering of wet-spells, and durations of the dry season. However, it

Map 2.4 Winter (December-January-February) Rainfall Totals (Millimeters) in 2007–08 (Left Panel, Positive NAO) and 2009–10 (Right Panel, Negative NAO)

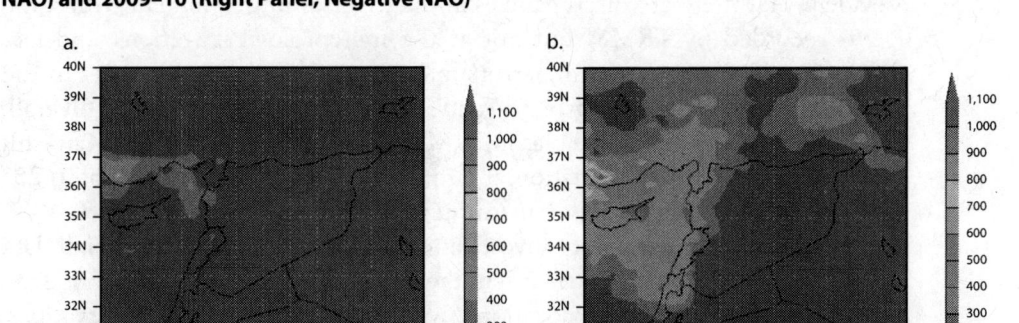

Source: The images and data used in this study were acquired using the GES-DISC Interactive Online Visualization and analysis Infrastructure (Giovanni) as part of the NASA's Goddard Earth Sciences (GES) Data and Information Services Center (DISC)

Map 2.5 Accumulated Rainfall Totals (Millimeters) from Daily TRMM for the Decade 1998–2007

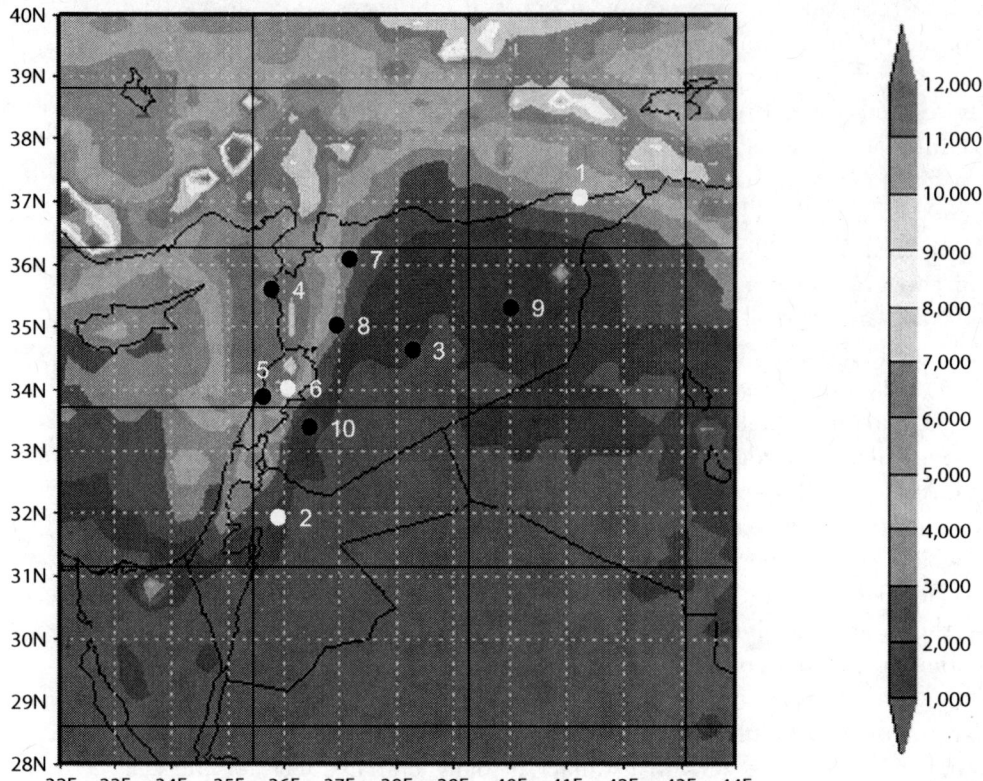

Source: The images and data used in this study were acquired using the GES-DISC Interactive Online Visualization and analysis Infrastructure (Giovanni) as part of the NASA's Goddard Earth Sciences (GES) Data and Information Services Center (DISC)
Note: The grid cells of the HadCM3 climate model and three test sites are shown as white dots. Other sites for which there are downscaling results are shown as red dots.
(1) Kamishli, Syria (2) Amman, Jordan (3) Palmyra, Syria (4) Latakia, Syria (5) Beirut, Lebanon (6) Kfardane, Lebanon (7) Aleppo, Syria (8) Hama, Syria (9) Deir Ezzor, Syria (10) Damascus, Syria

is evident that there are discrepancies in the exact timing of rainfall with some events recorded by TRMM but not at the meteorological stations, and *vice versa* (figure 2.2). Some timing errors may be attributed to differences in the recording interval used by ground- and satellite-based observations. Overall, there is a ~7 percent negative bias in TRMM rainfall accumulations at Kamishli and ~35 percent underestimation at Amman. Differences between the 0.25° and 1° aggregations are smaller at Amman than Kamishli (figure 2.2).

From the above preliminary investigation of TRMM and station rainfall data it is concluded that the former offers the potential for hydrological modeling and impact assessment at locations without meteorological observations. However, further work is needed to establish robust scaling rules to correct for biases. These could be based on local physiographic information (such as elevation, latitude, longitude, distance, etc.) provided that a larger number of sites can be evaluated across the region.

Figure 2.1 Observed and Re-Scaled TRMM Daily Rainfall at Kamishli (Upper) and Amman (Lower)

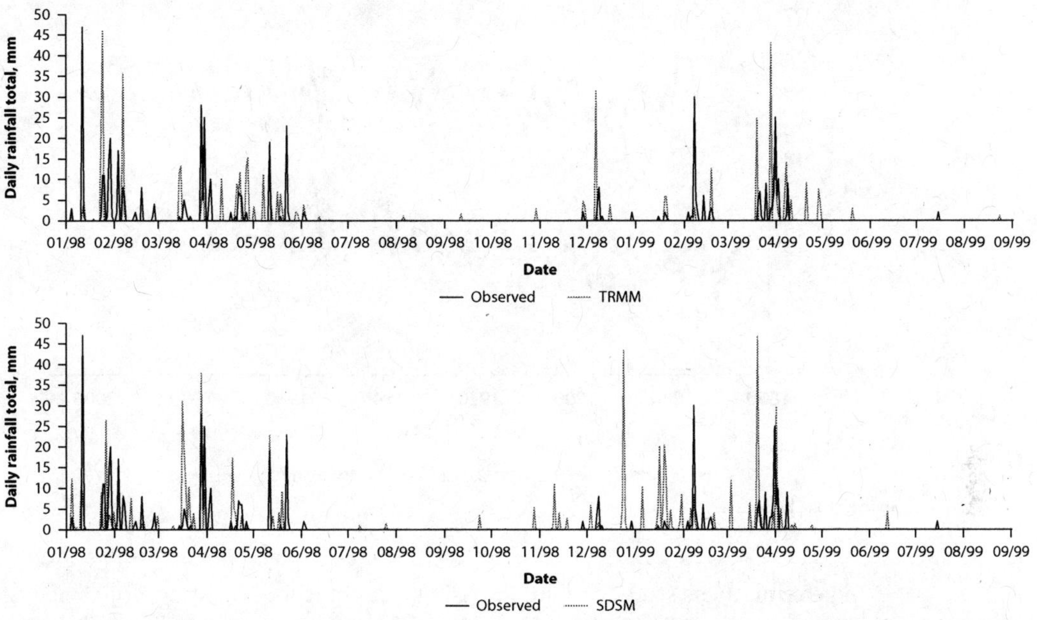

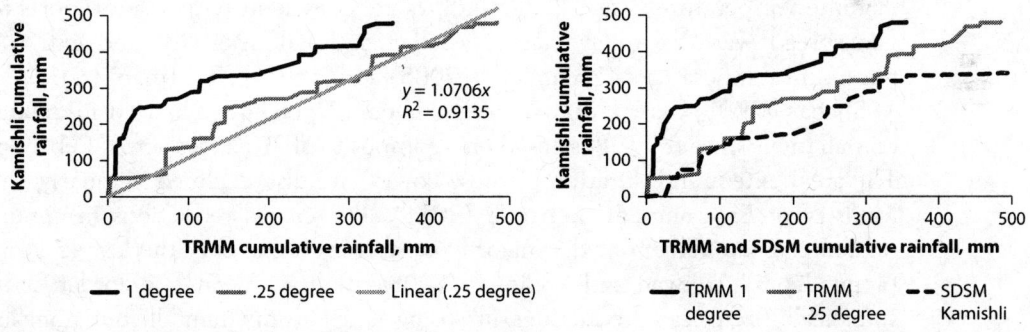

Source: World Bank data.

Figure 2.2 Observed and TRMM Rainfall Totals at Kamishli (Left) and Amman (Right Panel) 1998–99

Source: World Bank data.

Observed Climate Trends

Long-term, homogeneous observations of rainfall and temperature are scarce in
the region, so there is a limited scope for evaluating trends in climate variables,
especially for extreme events. Nonetheless, Freiwan and Kadioğlu (2008a,
2008b) and the Ministry of Environment, Jordan (2009) examined monthly
precipitation and temperature records with at least 30 years of data at representative stations in Jordan (see appendix B). Shahin (2007) looked for trends in
monthly rainfall, temperature, and evaporation at sites across the Middle East.

Figure 2.3 Annual Mean Temperatures for Selected Stations in the HadCRUT3 Archive

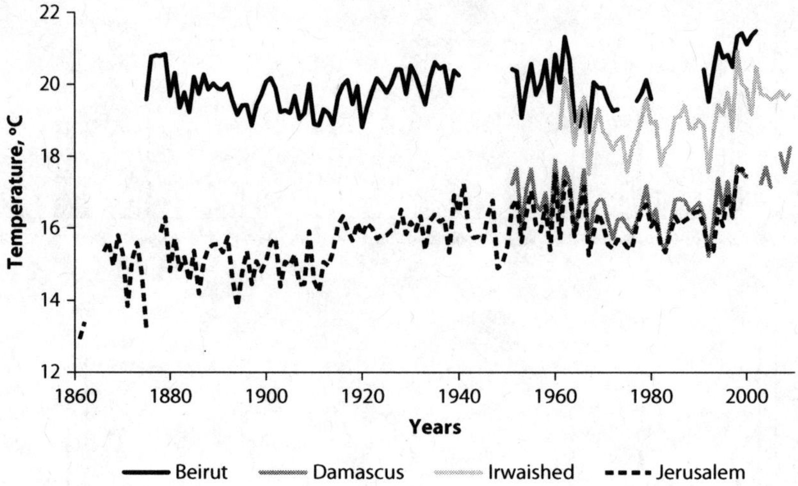

Source: World Bank data.

In general, these studies highlight rising temperatures, most notably since the 1970s (see figure 2.3). The greatest warming has occurred in summer nighttime temperatures. For example, summer minimum temperatures have risen by 3.6° per century and summer maxima by 1.8° per century at Kimishli. With little or no change in daytime temperatures there has been an overall decrease in the daytime temperature range. These findings are consistent with other reports of widespread warming of average (Nasrallah and Balling 1993) and extreme minimum temperatures (Zhang et al. 2005) across the region (map 2.6).

Shaban (2009) asserts that there has been 12 percent reduction in annual rainfall totals since the 1980s based on a composite of 70 gauges across Lebanon. The areal extent and duration of snow cover have also declined. Similarly, the Ministry of Environment, Lebanon (2011) also reports reductions in annual rainfall of 5–20 percent at the majority of sites in Jordan over the last 45 years (appendix B). Freiwan and Kadioğlu (2008a) find a few sites in Jordan with statistically significant reductions in spring and autumn rainfall, but none in winter. It is speculated that these changes may have resulted from a shift in cyclone track over the last 50 years. However, it is apparent that these trends are small compared with the large inter-annual variability in rainfall totals (figure 2.4).

Trends in daily precipitation indices for the Middle East, including the number of days with precipitation, the average precipitation intensity, and maximum daily precipitation amounts, are generally weak and do not show spatial coherence (Zhang et al. 2005). However, there is some local evidence of increasing frequency of heavy rainstorms and extreme floods in Israel (Greenbaum, Schwartz, and Bergman 2010). Evaporation has decreased at many sites across Jordan in line with cloudier conditions (appendix B).

Map 2.6 Trend in Annual Cool Days (TX10p), Cool Nights (TN10p), Hot Days (TX90p), and Hot Nights (TN90p) for the Periods 1950–2003 and 1970–2003

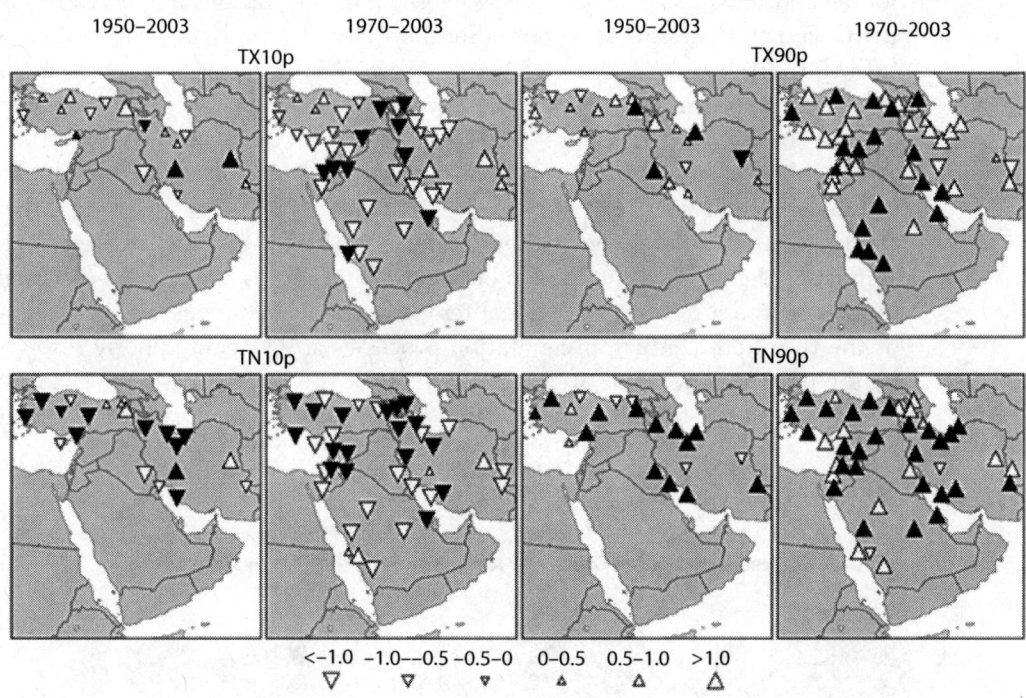

Source: Zhang et al. 2005, Figure 5 and Figure 6.
Note: Upward triangles represent increasing trends, downward triangles decreasing trends. Triangle size indicates the magnitudes of trend (in percentage per decade). Solid triangles represent trends significant at the 5 percent level.

Figure 2.4 Annual Precipitation Totals for Selected Sites in the GHCN Archive

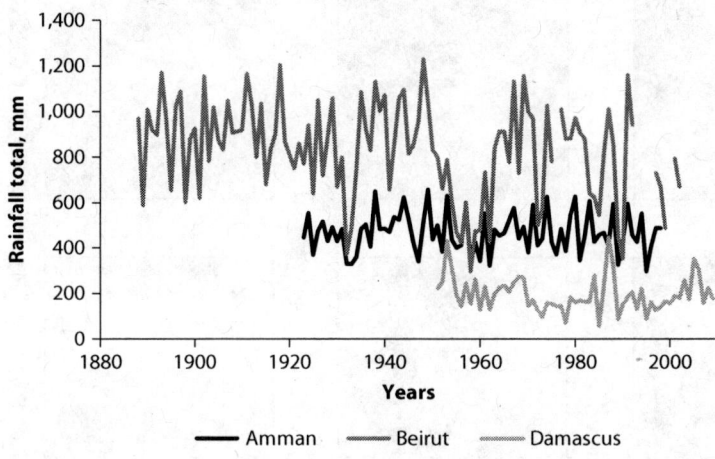

Source: World Bank data.

Climate Model Projections

Jordan and Lebanon lie within three overlapping sub-regions employed by the IPCC Fourth Assessment Report: Southern Europe and the Mediterranean (SEM), Central Asia (CAS) and North Africa (NAF). For the purpose of this report, the African domain provides most complete coverage of the areas of interest (map 2.7).

As noted above, the region exhibits high natural inter-annual climate variability and strong climatic gradients due to the complex interplay of topography, inland seas and marine influences. All of these features mean that simulating the regional climate can be problematic (Evans et al. 2004). Although models reproduce observed regional temperature changes over the second half of the twentieth century, precipitation processes are generally poorly resolved by the present generation of climate models for the Middle East. For example, one evaluation of the IPCC climate model ensemble found that only four out

Map 2.7 Temperature and Precipitation Changes Over Africa from the MMD-A1B Simulations

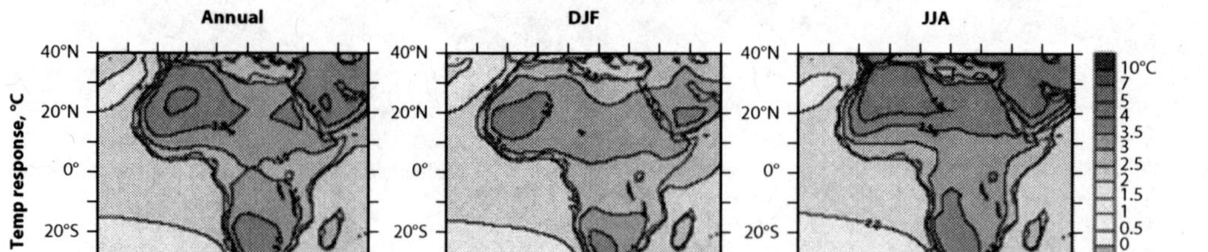

Source: Christensen et. (2007) Figure 11.2.
Notes: Top row: Annual mean, winter (DJF) and summer (JJA) temperature change between 1980–99 and 2080–99 averaged over 21 models. Middle row: same as top, but for fractional change in precipitation. Bottom row: number of models out of 21 that project increases in precipitation.

of 18 models performed better than the long-term mean estimate of rainfall (Black 2009). Furthermore, even under observed climate conditions, higher-resolution RCMs such as RegCM3 tend to overestimate rainfall totals in areas of high relief (Pal et al. 2007).

Despite these acknowledged limitations, a consistent picture is beginning to emerge for the Middle East from a growing number of climate model experiments. One multi-model ensemble predicts a mean temperature increase of ~1.4°C by mid-century (2045–54) and over 2.5°C by the end of the century (2090–99) (Evans 2009). The largest temperature increases are expected to occur away from water bodies in the vicinity of east Syria and Iraq. Furthermore, some RCMs suggest that the temperature increases will be greatest in summer (Önol and Semazzi 2009).

Predicted changes in precipitation are less certain, although unlike many other regions (see Figure 2.1) there is a strong consensus about the direction of the trend in the northern part of the domain towards drier conditions. The most significant reductions in precipitation are projected for an area covering the Eastern Mediterranean, including Lebanon and Jordan. Here, annual precipitation could decline by more than 100 millimeters per year compared with the present (Evans 2009).

Aslightly different multi-model ensemble forced by the SRES A1B emissions scenario shows reductions of up to 25 percent in the same region by the end of the century (Kim and Byun 2009). The study also suggests that drought durations could extend by 60 percent and that extreme droughts could occur more frequently. The authors predict that the Arabian Desert will become more arid and expand northward into Syria and its vicinity (Kim and Byun 2009, 149). Comparable results were obtained from RegCM3 under SRES A2 emissions. This experiment showed reductions in winter precipitation in the range 24–32 percent over Jordan and Lebanon (Gao and Giorgi 2008; Önol and Semazzi 2009). However, it is noted that autumn precipitation over the source areas of the Euphrates and Tigris Rivers could increase by 48 percent, thereby helping to compensate for the downstream winter deficit.

Several studies consider the underlying causes of projected changes in precipitation in the Middle East. It is generally accepted that global climate change will cause Northern Hemisphere storm tracks to move polewards and hence weaken the strength of the Mediterranean storm track (Bengtsson, Hodges, and Roeckner 2006). This is expected to reduce the number of cyclones crossing the Mediterranean (Giorgi and Lionello 2008; Lionello and Giorgi 2007) and hence the frequency and duration of rainfall events over Jordan, Israel, and the West Bank and Gaza (Black 2009). However, one study based on a single GCM found that the reduction of cyclones was compensated by an increase of polar intrusions, resulting in no net change to winter rainfall in the Eastern Mediterranean by the 2080s (Raible et al. 2010; see also Krichak, Alpert, and Kunin 2010). Another suggests that rainfall over the Anatolian Peninsula could increase with higher sea surface temperatures in the Mediterranean (Bozkurt and Sen 2009).

Statistical Downscaling Model

The preceding discussion of climate change projections for the Middle East referred mainly to the outputs of dynamical downscaling, namely Regional Climate Models (RCMs). These tools provide information at resolutions of 10–50 kilometers. For finer scale applications, the alternative is statistical downscaling. For an overview of the strengths and limitations of each technique the reader is referred to Wilby and Fowler (2010); for a discussion of the use of climate model scenarios in development and adaptation planning see Wilby et al. (2009).

From here on the Statistical DownScaling Model (SDSM) (Wilby, Dawson, and Barrow 2002) is used to construct exploratory scenarios of daily mean temperature and daily precipitation totals at selected sites for the period 1961–2099 under SRES A2 and B2 emissions. As with other statistical downscaling methods, SDSM derives physically sensible empirical relationships between the local variable(s) of interest (such as daily precipitation in Amman) and large-scale atmospheric predictors (such as sea level pressure, vorticity, geopotential heights, and humidity over the eastern Mediterranean) supplied by gridded re-analyses (map 2.5). These relationships are assumed to be valid under future greenhouse gas forcing.

Statistical downscaling requires observed daily mean temperature and rainfall data at the site(s) of interest for any years within the calibration period of 1961–2000. SDSM is then calibrated using the available observations and large-scale predictor variables for 1961–2000 sourced from the National Center for Environmental Prediction (NCEP) re-analysis (appendix C). These predictors are obtained in SDSM format from the Canadian Climate Change Scenarios Network (CCCSN)[8] for grid-cells over or adjacent to the target sites. All data must be quality assured and any anomalous outliers removed.

The following results illustrate the capability of SDSM at simulating observed temperature and precipitation regimes of Amman (Jordan) and Kfardane (Lebanon), recognizing that regional climate modeling for mountainous, semi-arid, and arid continental climates is highly challenging (Evans, Smith, and Oglesby 2004). Downscaling model skill was assessed using a range of metrics, including seasonal means, daily time series, and distribution attributes. These provide reassurance (but by no means absolute confidence) in the robustness of climate change scenarios generated by the same model variants.

Model Testing and Evaluation

SDSM faithfully reproduces the temperature regime at Amman across a range of time-scales: day-to-day, intra-annual, and inter-annual variability are all simulated convincingly (figures 2.5 and 2.6). This suggests that the suite of predictor variables used for the downscaling has captured the underlying drivers of temperature at this site (appendix C). The apparent discrepancy between observed and downscaled annual maximum temperatures (figure 2.6, right panel) is thought to reflect both downscaling model bias and the fact that there are 18 years of data missing from the 40 year observational record. Therefore, the curve for observations reflects only a sample of the years hindcast by the downscaling.

Figure 2.5 Downscaled and Observed Daily Mean Temperature at Amman for 1999–2000

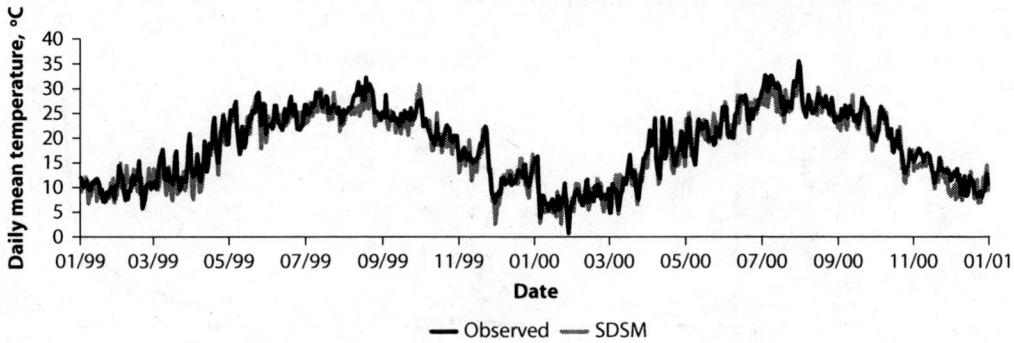

Source: World Bank data.

Figure 2.6 Hindcasts of Winter Mean Temperatures (Left Panel) and Estimated Return Periods for Daily Mean Temperatures (Right Panel) at Amman

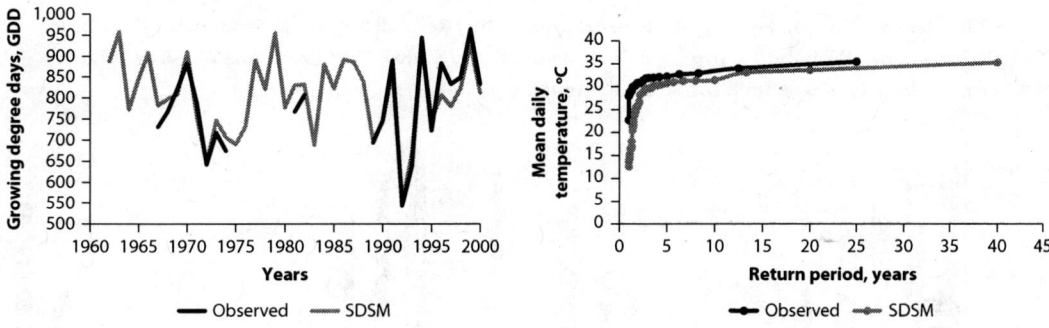

Source: World Bank data.

As with temperature downscaling, it is desirable that variability in precipitation amounts is captured realistically at all time-scales, from daily sequences through to decadal trends. However, the predictability (that is, amount of explained variance) in local precipitation given large-scale atmospheric properties is much lower for precipitation than temperature (see appendix C). SDSM addresses this unexplained variance by adding "white noise" to predictions so that their overall distribution better matches observations. In the case of rainfall, this "random" component can be large, such that the exact sequencing of daily amounts would not be expected to exactly match reality.

A cleaner test of the model capability is to compare probability distributions of observed and downscaled daily rainfall amounts for a common reference period. Overall, SDSM yields realistic distributions of wet-day amounts at both demonstration sites for the period 1961–2000, although the magnitude of the heaviest events is under-estimated (figure 2.7).

The downscaling results for Amman are not very convincing (figure 2.8). The model reproduces rainy season (October to March) precipitation totals but

Figure 2.7 Observed (Black Line) and Downscaled (Grey Line) Distributions of Wet-Day Totals at Amman for the Period 1961–2000.

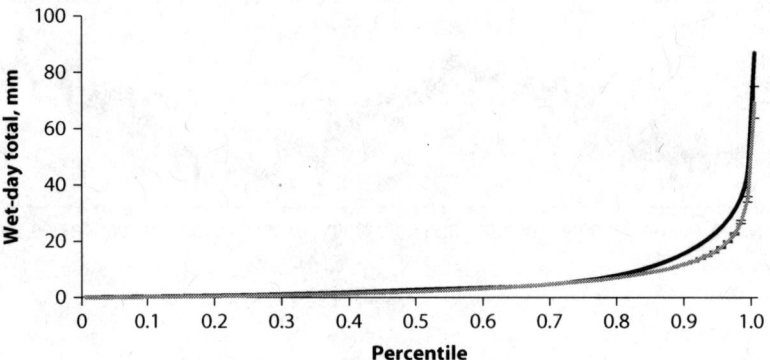

Source: World Bank data.
Note: T-bars show standard errors of the ensemble.

Figure 2.8 Observed (Black) and Downscaled (Grey) Monthly Rainfall Metrics at Amman for the Period 1961–2000: Wet-day Probability (Top Left), Total Rainfall (Top Right), 95th Percentile Wet-Day Total (Bottom Left) and Mean Dry-Spell Duration (Bottom Right)

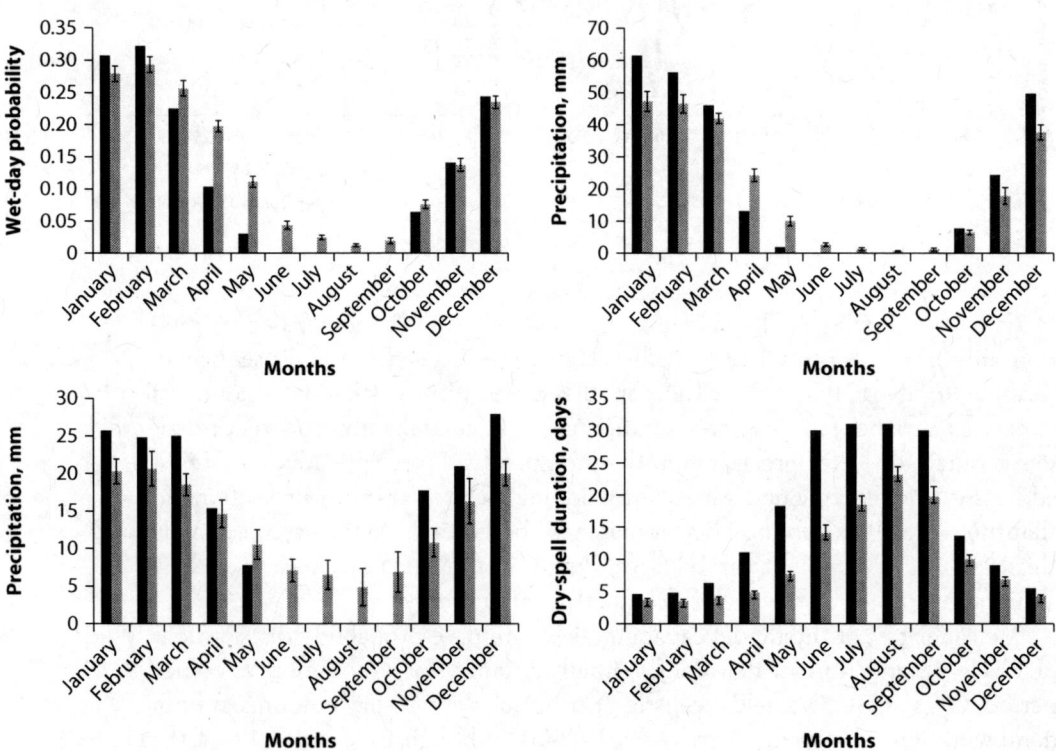

Source: World Bank data.

over-estimates wet-day occurrence and amounts in the summer. This in turn causes the downscaled dry-spell duration to be too short in the summer. These model biases arise because there are insufficient data to calibrate the model on a month-by-month or even seasonal basis. In other words, a single statistical model is being used to downscale the daily rainfall series regardless of the time of year. Note also that abrupt changes in annual rainfall totals and the number of rain days have been reported for Amman, albeit for the mid-1950s, bringing into question the homogeneity of the record (see Smadi and Zghoul 2006). Further software development is needed to enable model calibration on a monthly basis even when there are no data for particular seasons (for example, when there are no rain days in summer). This would enable more faithful reproduction of the intra-annual rainfall regime in arid climates.

Infilling and Hindcasting Missing Data

As noted previously, long, homogeneous daily meteorological records are relatively rare in the Middle East. However, where there are sufficient high quality data to calibrate SDSM, there is scope for infilling and hindcasting missing data using identified relationships between large-scale predictor variables and local weather. For example, figure 2.9 shows observed and downscaled daily mean temperatures at Kfardane, Lebanon, where records began in 1994 and run to present. Using all available data for the period 1994–2000, it is possible to hindcast daily temperatures and derived metrics (such as the winter growing degree days) for periods that pre-date observations. Similarly, days with missing data (as in May 1994) can be estimated using the same statistical model. Where multidecadal series have been reconstructed it is also possible to estimate historic trends. For example, the SDSM hindcast of precipitation totals at Kfardane suggests a long-term decline in annual rainfall (~125 millimeters) over the period 1961–2000 (figure 2.10).

Figure 2.9 Observed and SDSM Hindcast Daily Mean Temperature (Left) and Winter Growing Degree Days (Right) at Kfardane, Lebanon

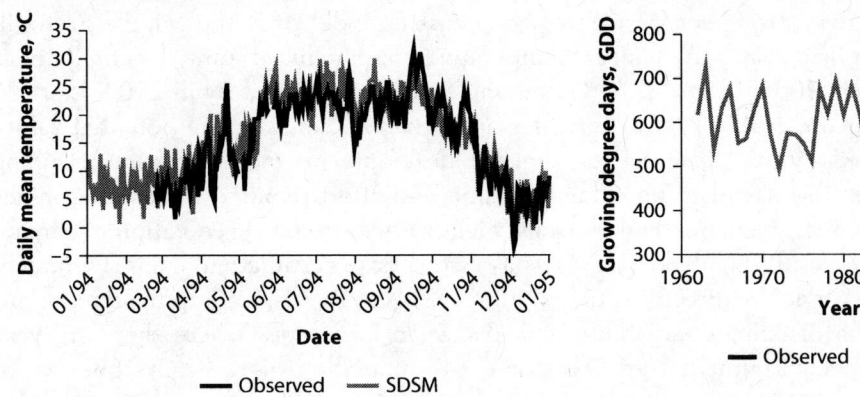

Source: World Bank data.
Note: SDSM = Statistical DownScaling Model.

Figure 2.10 Observed and SDSM Hindcast Wet-day Precipitation Amount Distribution (Left) and Annual Precipitation Totals (Right) at Kfardane, Lebanon

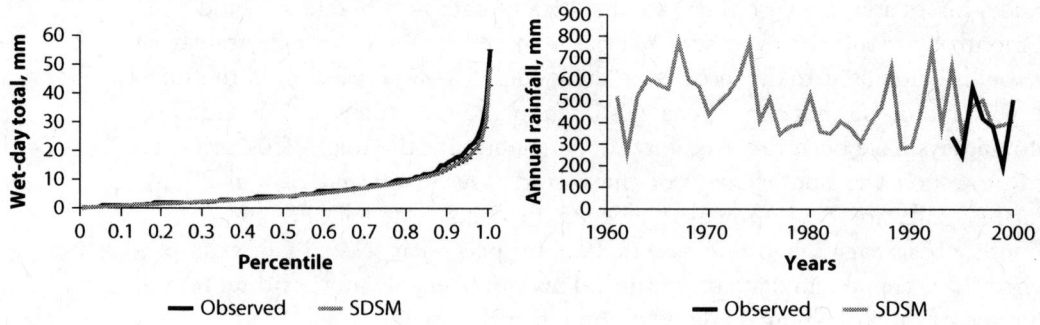

Source: World Bank data.

Notes: SDSM = Statistical DownScaling Model. Observed (black line) and downscaled (gray line) distributors of wet-day rainfall amounts at Kfardane for the period 1961–2000. T-bars show the standard error of the model estimates.

Overall Assessment

Comparisons of observed and downscaled statistics would not be expected to match exactly because of missing data in the reference period (in the case of precipitation for Amman this is 2 percent of the record). It is also recognized that mean statistics do not adequately represent semi-arid and arid climates because of the distorting influence of highly localized, extreme precipitation events. Even so, the diagnostics presented in figures 2.13 to 2.16 suggest that the downscaling captures many of the key features of the temperature and precipitation regimes given only information about large-scale atmospheric conditions over the region. On this basis, the model was judged "fit for purpose," so temperature and precipitation scenarios were downscaled for each station using information supplied by HadCM3 under the SRES A2 and B2 emissions scenario.

Statistical Downscaling Experiments

There is strong consensus amongst climate models that the Middle East will experience increased mean temperatures and reduced annual rainfall totals (Black 2009; Evans 2009; Giorgi and Bi 2005; Kim and Byun 2009; Lionello and Giorgi 2007). Therefore, it is important to consider the potential value-added by statistical downscaling to development and adaptation planning given the attendant uncertainties, time, and effort required to construct high-resolution scenarios. Four reasons might be ventured: (1) generation of climate scenarios for highly localized risk assessments; (2) estimation of climate indices that cannot be directly obtained from GCMs or RCMs; (3) improved representation of climate variability and change in landscapes where there are very steep environmental gradients; and (4) simulation of realistic sequences of daily weather for evaluation of adaptation options under present and future climate conditions.

This section describes a few experiments that illustrate the potential of statistical downscaling for estimating climate indices that are meaningful to decision making (that is, justification (2) above). Following Evans (2009) three diagnostics are chosen: (1) frequency of exceedance of 200 millimeters per year (the widely accepted threshold for rain-fed agriculture in the region); (2) winter growing degree days (GDD); and (3) the length of dry-spells as a measure of drought persistence (given that in the future there are expected to be fewer cyclones in the Eastern Mediterranean). Summary statistics are provided for mean temperature and precipitation changes downscaled at selected sites.

Mean Temperature and Precipitation

The range of downscaled mean annual temperatures changes of +1.3°C to +2.7°C by the 2050s and +1.9°C to +4.6°C by the 2080s—depending on emissions scenario and location (table 2.1)—bracket the results of Evans' multimodel ensemble (Evans 2009). However, annual statistics conceal large intra-annual variations in the rate of warming. Most rapid warming occurs in spring and summer. For example, by the 2080s under SRES A2 emissions, May temperature changes could be +6.5°C at Kfardane (figure 2.11). Much less warming is projected in summer for the coastal zone (Beirut).

Again, in line with Evans (2009), the downscaled scenarios have significant reductions in annual precipitation totals at all test sites (table 2.2 and figure 2.12). Depending on the emissions scenario and location, precipitation could decrease by 14–51 percent by the 2080s.[9]

Frequency of Annual Rainfall Less Than 200 Millimeters

Evans and Geerken (2004) suggest that the limit to rain-fed agriculture lies very close to the 200 millimeters per year isohyets. At present, the mean annual rainfall total at Amman is ~260 millimeters (1961–2000) so the downscaled reductions in table 2.2 suggest that rain-fed agriculture would not be viable in an average year of the 2080s. However, due to natural variability combined with the climate change signal, the practice could be uneconomic much sooner. Transient runs of the downscaled rainfall totals for the period 1961–2099 were used to investigate the growing risk of annual rainfall less than 200 millimeters through time (figure 2.13).

Table 2.1 Changes (°C) in Mean Annual Temperature Downscaled to Selected Sites from HadCM3 under SRES A2 and B2 Emissions

Site	2020s		2050s		2080s	
2.1	**A2**	**B2**	**A2**	**B2**	**A2**	**B2**
Amman	+1.3	+1.5	+2.3	+2.1	+4.0	+3.0
Beirut	+0.8	+0.9	+1.4	+1.3	+2.5	+1.9
Kfardane	+1.2	+1.5	+2.2	+1.9	+3.8	+2.9

Source: World Bank data.
Note: See appendix D for additional sites and seasonal breakdown of changes.

Figure 2.11 Changes in Monthly Mean Temperatures (°C) Downscaled from HadCM3 to Amman and Kfardane under SRES A2 (Left Panels) and B2 (Right Panels) Emissions.

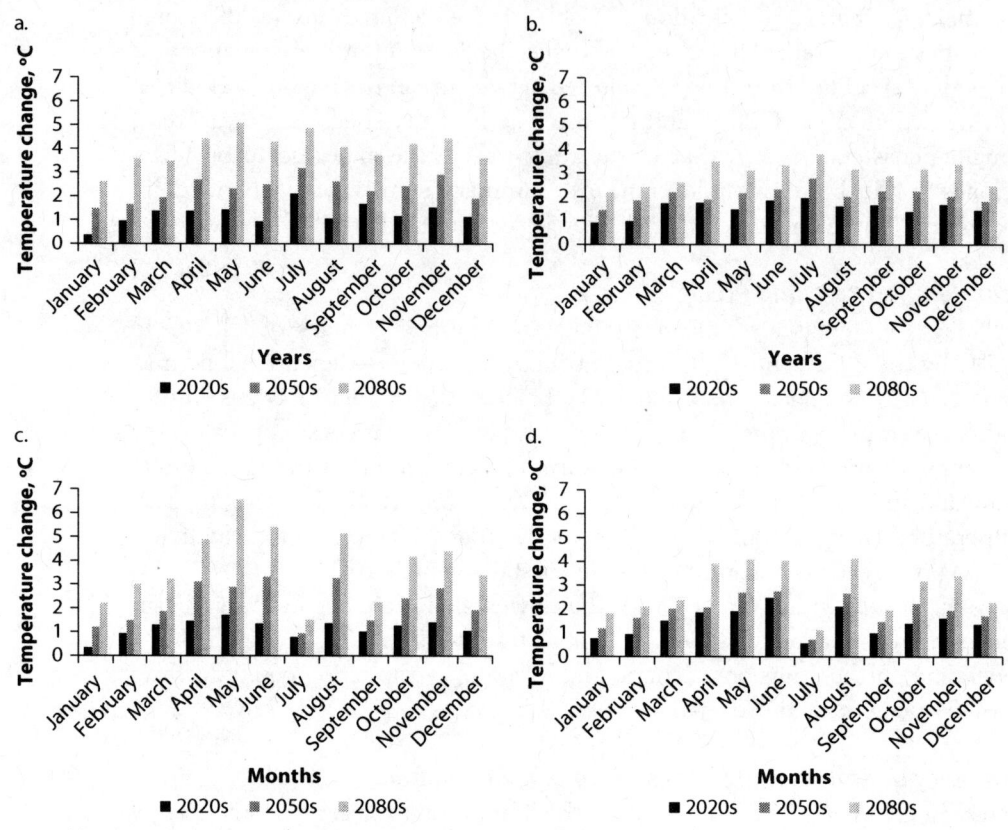

Source: World Bank data.

Table 2.2 Absolute Changes in Mean Annual Precipitation Totals (Millimeter) Downscaled to Selected Sites from HadCM3 under SRES A2 and B2 Emissions

Site	2020s		2050s		2080s	
	A2	B2	A2	B2	A2	B2
Amman	−36 (−18)	−49 (−23)	−60 (−29)	−52 (−25)	−105 (−51)	−77 (−37)
Beirut	−47 (−7)	−76 (−11)	−74 (−11)	−52 (−8)	−160 (−24)	−96 (−14)
Kfardane	−60 (−13)	−87 (−19)	−113 (−24)	−107 (−23)	−208 (−44)	−148 (−31)

Source: World Bank data.
Note: Percent changes are shown in brackets. See appendix E for additional sites and seasonal breakdown of changes.

The coefficient of variation of observed annual totals is 38 percent and there has historically been a one in three chance of amounts less than 200 millimeters in any given year. However, there is evidence that the risk of dry years increased during 1961–2000 and this trend is expected to continue in the future under both emissions scenarios (figure 2.13, right panel). Under

Figure 2.12 Changes (Millimeters) in Monthly Precipitation Totals Downscaled from HadCM3 to Amman and Kfardane under SRES A2 (Left Panels) and B2 (Right Panels) Emissions

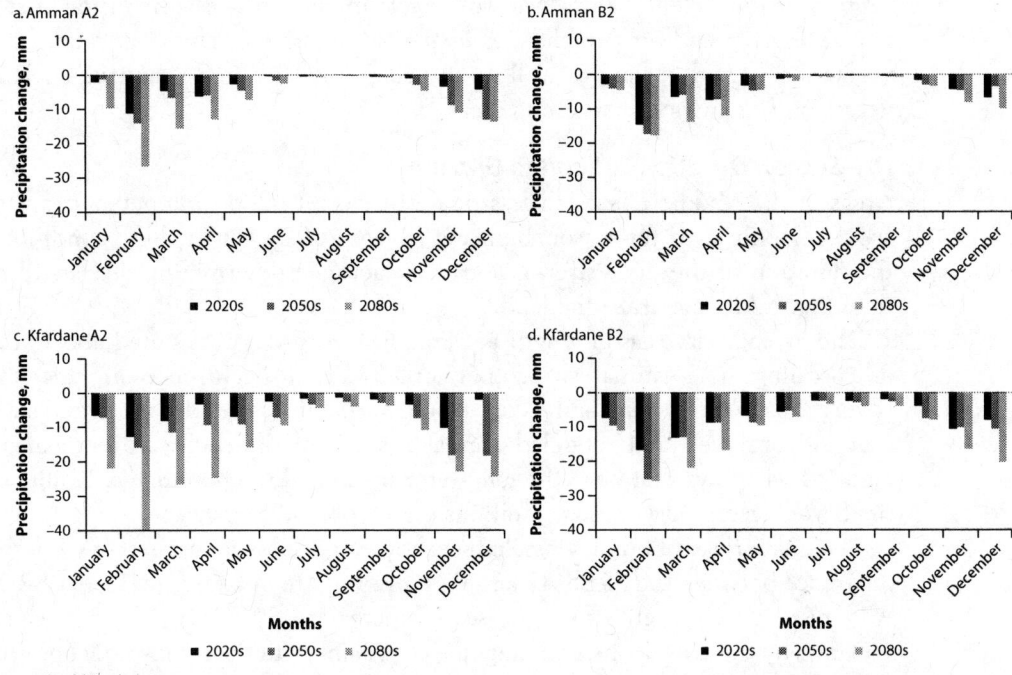

Source: World Bank data.

Figure 2.13 Observed (OBS) and Downscaled (SRES A2 and B2) Annual Rainfall Totals (Left Panel) with 30-Year Moving Averages of the Chance of Years with Totals below 200 Millimeters (Right Panel) at Amman 1961–2099

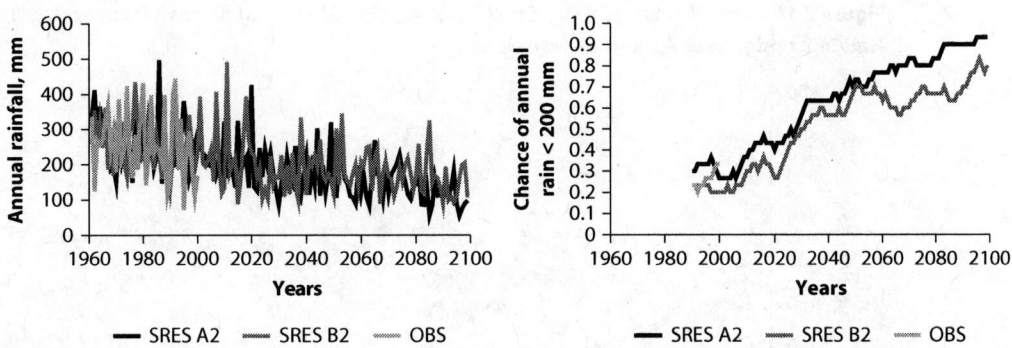

Source: World Bank data.
Note: The SRES A2 and B2 series were scaled to correct for biases in the HadCM3 predictors and downscaling scheme to match observed totals over the period 1961–2000.

the illustrative HadCM3-A2 scenario, the chance of a dry year becomes one in two (that is, 50 percent) by the late 2020s. The dry-year frequencies of the two emission scenarios begin to diverge beyond the 2050s. By the 2040s two years in three could have less than 200 millimeters.

These findings are consistent with previous studies showing that the 200 millimeters isohyet could move ~75 kilometers northward by the end of the century (Evans 2009). However, the exact impact on the agricultural sector would also depend on any changes in the timing of the rainfall in relation to vegetation growth stages, as well as any adaptation of local practices, such as extension of rainwater harvesting areas.

Dry-Season Duration as Limit to Grazing

Areas of the Middle East with less than but close to 200 millimeters per year are commonly used for seasonal grazing (Evans 2009). Under these conditions, the duration of the dry season is a critical factor in determining the length of time that herds can graze the land.

The average dry-season length at Amman is 175 days. As noted above, the downscaling over-estimates summer precipitation occurrence at this site (figure 2.8) but this bias can be corrected by applying a wet-day threshold such that the observed and modeled annual maximum dry-spells are equivalent. Figure 2.14 shows that both the long-term mean and inter-annual variability of the bias-corrected dry-season durations match observations.

Looking forward, both downscaled scenarios suggest longer dry-seasons, but the difference between the SRES A2 and B2 emissions are insignificant. Under SRES A2 emissions the length of the dry-season increases ~30 days by the 2050s and ~63 days by the 2080s. These findings are consistent with the multi-model projections of Evans (2009) which showed an increase of ~2 months by the end of the twenty-first century. Both sets of results point to a future in which there may be increased reliance on imported fodder and/or decreased herd sizes in the region.

Figure 2.14 Annual Maximum Dry-Spell Duration (Dry Season) at Amman Downscaled from HadCM3 under SRES A2 and B2 Emissions

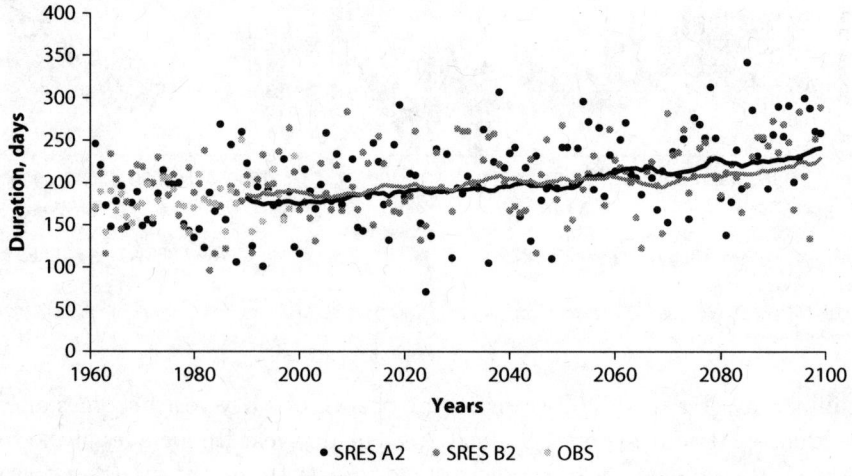

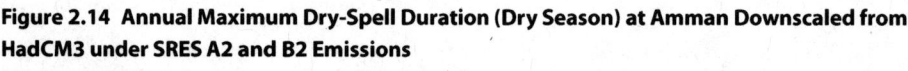

Source: World Bank data.
Note: A wet-day threshold of 4.5 millimeters was used to adjust for known biases in the HadCM3-SDSM downscaling.

Downscaling to Agro-Ecological Zones

Across the region, there are complex relationships between projected changes in mean annual temperature and precipitation totals that depend of the physiographic context (figure 2.15). For instance, the amount of warming tends to increase with elevation and distance from the Mediterranean such that future rates of warming are less in the coastal zone (Lattakia, AEZ 1a) than in the desert interior (Palmyra, AEZ 5). This finding is consistent with other research showing more rapid warming at higher altitudes than at lowland (Beniston 2003; Pepin and Losleben 2002) and/or coastal sites (Ragab and Prudhomme 2002).

In line with earlier work, the largest reductions in rainfall are projected for Lebanon. Away from the coastal zone, elevation also exerts an influence with higher altitude sites (such as Amman, equivalent to AEZ ~2/3, and Kfardane equivalent to AEZ 1b) experiencing large absolute reductions in rainfall. This is because the likelihood of rising air and hence orographic precipitation increases when the winter and spring moisture flux is from west to east (Evans, Smith, and Oglesby 2004). Therefore, the projected poleward shift in zonal airflow (Bengtsson, Hodges, and Roeckner 2006) would be expected to reduce orographic rainfall in the region's coastal zone and mountain ranges.

Subject to availability of daily meteorological data, these generalizations could be examined in further detail using a larger set of representative stations

Figure 2.15 Changes in Annual Mean Temperature (Left Column) and Precipitation Totals (Right Column) in Relation to the Site Elevation (Upper Panels) and Longitude (Lower Panels)

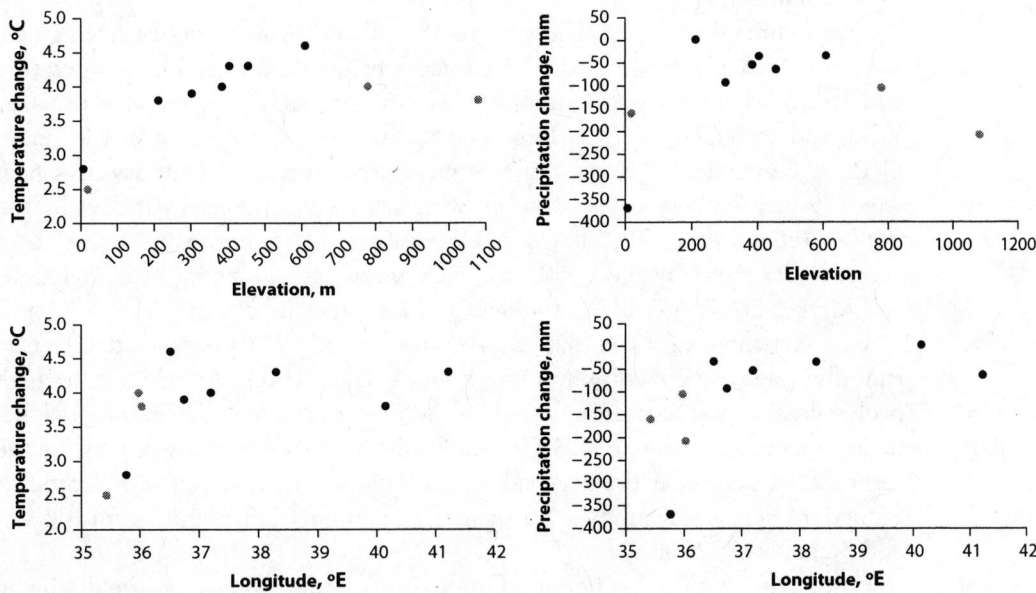

Source: World Bank data.

Note: All scenarios were downscaled from HadCM3 under SRES A2 emissions for the 2080s. Black symbols denote Syrian stations; red symbols are for sites in Lebanon (Beirut and Kfardane) and Jordan (Amman).

distributed across the region. Downscaling could also be performed using an ensemble of climate models to better characterize the spread of projected changes. However, given the strong consensus amongst climate models about the projected changes in regional temperature and precipitation, there may be less demanding ways of achieving the project objectives. For example, pattern-scaling techniques could be applied to baseline climate data (such as the TRMM precipitation distribution in map 2.5) thereby enabling fully distributed impacts modeling for a range of climate scenarios. An alternative strategy would be to downscale to representative sites in each AEZ and to use a manageable number of "marker" scenarios to evaluate the performance of different adaptation strategies at each location. Much insight can also be gained from relatively straightforward sensitivity of sector models.

Impacts of Climate Change

There have been a growing number of studies looking specifically at climate change impacts on the Middle East (see collection of papers in Zereini and Hötzl 2008). Potential impacts are expected to be particularly acute in the region on account of the inherent variability of the climate, rapid population growth, water supply-demand mismatch, and heavy reliance on climate-vulnerable sectors such as agriculture (see box 2.1). After a review of regional impacts, the following sections provide a summary of the current situation in each country and a synthesis of studies relevant to water and agricultural sectors.

Regional Impacts

Potential climate change impacts for the Middle East can be drawn from global assessments. A composite index of climate change (comprised of temperature and precipitation extremes) indicates mid-range impacts for the region (Baettig, Wild, and Imboden 2007). More specifically, a multi-model ensemble under SRES A1B emissions shows significant increases in the number of dry days, heat waves, warm nights and length of growing season, and fewer frost days by the 2080s (Tebaldi et al. 2006). The Middle East also emerges as a "hot spot" of severe water stress by the 2050s in several global assessments (Alcamo, Flörke, and Märker 2007; Arnell 2004; Ragab and Prudhomme 2002).

One ensemble of 12 climate models projects a 10–30 percent decrease in runoff by the year 2050 (Milly, Dunne, and Vecchia 2005). Another shows high probability of increased drought risk in the Eastern Mediterranean under global mean warming greater than 3°C (Scholze et al. 2006). An assessment of the response of perennial drainage networks to lower rainfall showed significant reductions across a swathe of North Africa including Sudan and Egypt (de Wit and Stankiewicz 2006).

Other studies consider potential impacts of climate change on the discharge regimes of the region's transboundary aquifer and river systems. For example, the mean annual flow of the Upper Jordan River system is estimated to be

~1,300 million cubic meters of which ~47 percent is abstracted by Israel, 22 percent by Jordan, 16 percent by Syria, and 2 percent by Lebanon, leaving only 13 percent of the flow to enter the Dead Sea (Al-Weshah 2000). Samuels et al. (2010) coupled a dynamical downscaling model (RegCM3) to a hydrological model (HYMKE) to evaluate the impact of climate change on the upper catchment of the River Jordan and its tributaries. With a 10 percent reduction in rainfall, base and surface flows decline by 10 percent and 17 percent respectively. However, the variance of flow increases, indicating greater variability and risk of extreme high flows (Samuels, Rimmer, and Alpert 2009).

Nohara et al. (2006) investigated future changes in discharge for 24 major rivers using the output of 19 climate models forced by the SRES A1B emissions scenario. Simulations of the present flow regime in the Euphrates were credible (although the modeled peak flow was too early and too low compared with observations). Simulations of the future show a 38 percent reduction in annual mean discharge by the 2090s. Other studies report reductions of 10–25 percent in the runoff of the upper Euphrates by 2070 (Meslmani 2008; see also: Bozkurt et al. 2010). Changes of these magnitudes would clearly have a profound impact on irrigators who consume the largest portion of the available resources.

Lautze and Kirshen (2009) model water allocation between Israel and the West Bank and Gaza under prescribed scenarios for population growth, climate change and management position to 2025. Under the business-as-usual scenario the "environmental surplus" would decline by between 13 percent and 48 percent by 2025 depending on the water-sharing formula. Similarly, Mizyed (2009) evaluated water resource availability and agricultural water demand in the West Bank under three temperature (+2, +4 and +6°C) and two precipitation scenarios (no change and −16 percent) using the Penman-Monteith equation for evapotranspiration and GIS mapping. Temperature changes alone were found to increase agricultural water demand by up to 17 percent and reduce annual groundwater recharge by 21 percent. When combined with precipitation reductions, groundwater recharge is diminished by up to 50 percent compared with the present.

Lebanon
Relative to other countries in the region, Lebanon is well endowed with renewable water resources of ~8,600 million cubic meters per year from 40 major rivers and more than 2000 springs yielding on average ~1,200 cubic meters per year per capita. However, the renewable freshwater has declined from ~1,900 cubic meters per year per capita in 1990 and the country is poised to fall into a water deficit within the next 10–15 years (Halwani 2009; Shaban 2009).

The growing imbalance of water supply and demand is partly due to trends in precipitation and snow cover over the last 40 years. Between the 1950s and 1980s precipitation in the Mount Lebanon basin dropped from 1,295 to

1,060 millimeters per year (Khair et al. 1994). Since the 1980s precipitation has decreased by 12 percent across Lebanon as a whole, whilst the average number and intensity of peak rainfalls has increased (Shaban 2009). The average discharge of Lebanese rivers is also falling (from 246 million cubic meters per year in 1965 to 186 million cubic meters per year in 2005), as is the number of springs (50–55 percent decrease) and volume of springflow (53 percent decrease) (Shaban 2009). Satellite measurements indicate that the area of dense snow cover in the Lebanese mountains has declined from 2,280 square kilometers before 1990, to an average of 1,925 square kilometers since (16 percent decrease). The average residence time of dense snow before melting has decreased too: from 110 days to less than 90 days over the same period (Shaban 2009).

The falling per capita water is also attributed to human factors including: growth in the urban population, lack of investment in infrastructure, unregulated sinking of private wells, overexploitation of aquifers, and resulting deterioration in the quality of groundwater. The coastal aquifers are particularly vulnerable to saline intrusion because of the combined effects of drought and abstraction beyond safe yields. It is estimated that there are over 10,000 wells tapping into the coastal aquifers of Beirut, and that chloride concentrations increased tenfold between 1970 and 1985 (Khair et al. 1994). Salinity concentrations greater than 5,000 milligram per litre are now detected in some public and private wells in the greater Beirut area, indicating a mixing of at least 10 percent seawater. Such concentrations render the water unsuitable for public supply and beyond the irreversible contamination limit (Saadeh 2008). Rates of saline intrusion could be exacerbated by any further decline in precipitation, increased evapotranspiration and/or sea level rise.

Hreiche, Najem, and Bocquillon (2007) modeled daily runoff and mean snow depth within the Nahr Ibrahim catchment for six climate change scenarios. The scenarios tested sensitivity to changes in rainfall amount, frequency, wet-spell duration, and length of rainy season, as well as the impact of a temperature increase of 2°C. Droughts were found to occur 15–30 days earlier, rainfall-generated floods replaced snowmelt events, and peak flows occurred two months earlier. The prescribed warming decreased the modeled depth of snow cover by ~50 percent.

Bou-Zeid and El-Fadel (2002) evaluated four climate model scenarios in terms of potential impacts on the water budget and soil moisture status in the Bekaa Valley and in Beirut. Evaporation was found to increase at both locations under all scenarios. The results suggest a possible increase in irrigation demand in the Bekaa Valley of up to 6 percent by the 2020s. Groundwater recharge was projected to decreases by up to 15 percent in Beirut over the same period. However, when averaged across Lebanon, climate-driven changes in renewable surface and groundwater are modest (<300 million cubic meters per year) in comparison to the projected impacts of population and economic growth (973 million cubic meters per year) by 2025. A range of adaptation options were considered, including the feasibility of exploiting the Chekka submarine springs[10] which lie within 1 kilometer of the coastline of northern Lebanon (Fleury, Bakalowicz, and de Marsily 2007).

Jordan

The annual average renewable water resources of Jordan are estimated to be ~800 million cubic meters, ranking the country as the seventh most water-poor in the world (Oroud 2008). Approximately 23 percent of the renewable surface water originates outside of the country. With an annual population growth rate of ~3.2 percent (in 2008), plus rapid expansion of the irrigated, and growing demand from industrial and tourism sectors, annual freshwater demands have risen from 1,100 million cubic meters in 1990, to 1,257 million cubic meters in 1998, and to 1,750 million cubic meters in 2004. Hence, the water status is already in significant overdraft and is highly vulnerable to the hotter and drier conditions expected with climate change.

Accordingly, several studies have investigated the sensitivity of water supplies and demand to incremental changes in temperature and rainfall. For example, Abdulla, Eshtawi, and Assaf (2009) find that the runoff of the Zarqa River Watershed is reduced by 1.2 percent per 1°C rise in mean annual temperature; groundwater recharge falls by three times this rate. Under the most extreme scenario (−20 percent annual rainfall combined with +3.5°C mean annual temperature) runoff decreases by ~23 percent. Oroud (2008) calculated that a 10 percent reduction in precipitation with a 2°C rise in temperature could reduce the water yield of the mountainous areas of Jordan by 40–60 percent. These could be plausible scenarios by the 2050s (according to the downscaling results for Amman presented in tables 2.1 and 2.2). Depleted flows in the River Jordan would accelerate the drop in water levels within the Dead Sea (Al-Weshah 2000) and reduce freshwater inflows to Lake Kinnerat (Samuels et al. 2010).

Abu-Taleb (2000) calculated the annual water deficit for Jordan as a whole, taking into account projected water use by all sectors for specified precipitation and temperature changes to 2020. Under a temperature rise of 4°C and a 20 percent decrease in precipitation, the projected deficit would be 1,020 million cubic meters per year, compared with a deficit of 408 million cubic meters per year under the no climate change scenario. Several options for reducing the deficit were tested, including: water pricing, conservation measures, water distribution network rehabilitation, stricter enforcement of metering, billing and revenue collection, and reallocation through volumetric constraints. Taken together, these measures could realize water savings of up to 566 million cubic meters. However, even an optimal combination of strategies would not produce enough water savings to offset the anticipated deficits. Under the best case scenario (that is no change in temperature, an increase in precipitation of 20 percent) Jordan would still need to invest in several of the water conservation measures.

Box 2.1

Biophysical Impacts of Climate Change on Agricultural Systems

It is widely expected that climate changes will impact agricultural systems in Jordan and Lebanon. This will happen through changes in temperature, moisture, and CO_2 levels, increased exposure to pests and diseases, and the interactions among all of these factors. It can be challenging to make concrete predictions of future impacts because of the complexity of agricultural systems and a lack of data on key environmental thresholds for many crops. Still, there is sufficient information available to identify general impacts and expected trends.

Effects from Temperature Changes

Plants often depend on temperature "cues" to trigger key developmental stages, such as germination, tillering and fruit ripening (Fuhrer 2003). Predicting specific crop responses to temperature is complex. The reasons for this are because: different species have different minimum and optimal temperatures for development; different processes occur at different times (for example, photosynthesis only occurs during light hours, while respiration occurs all day); and many of these processes are not related linearly to temperature (Gregory et al. 2009). For example, increased temperatures during the colder winter months in Jordan and Lebanon could mean that crops grown during these seasons mature sooner (Wilby 2010). While these changes could be beneficial in systems where the growing season is limited, in others, it could actually result in reduced yields. Temperature increases can accelerate a crop's development, which in turn can reduce the amount of time that crops like wheat or barley spend during the grain-filling stage (for example, producing grains), leading to smaller harvests (Khresat 2010). In addition, higher nighttime temperatures can increase overall crop respiration, potentially offsetting gains from increased day temperatures (Khresat 2010). Temperature increases can also affect the nutritional value of crops. High temperatures, pre- and post-harvest, can affect the quality of many fruit and vegetable crops, including reduced nutritional value as vitamin or antioxidant levels decrease and faster ripening and softening occurs (Moretti et al. 2009).

Effects from Precipitation Change

Drought stress occurs as a combination of two factors: when plants cannot access sufficient water through their roots (for example, if soil moisture levels are low) and when water losses are too high from transpiration (the loss of water through the stomata in leaves), which occurs if air temperatures are high or humidity levels are low. These two conditions often occur in semi-arid climates like those of Jordan and Lebanon, and are consistent with the predicted climatic changes in both countries (Reddy, Chaitanya, and Vivekanandan 2004).

The ability of plants to integrate carbon from the atmosphere during photosynthesis is decreased by water stress through drought. Therefore, decreasing water availability directly decreases a plant's capacity to grow. Reductions in vegetative growth, especially the growth of new shoots and leaves, are commonly seen in drought-stressed plants (Chaves and Oliveira 2004; Mahajan and Tuteja 2005). In severe situations, drought stress disrupts plant cell

(box continues on next page)

Box 2.1 Biophysical Impacts of Climate Change on Agricultural Systems *(continued)*

membranes and can inhibit enzyme functioning, disrupting essential metabolic plant processes such as photosynthesis and respiration. The precise effects of water stress on plants depend on the timing, intensity, and duration of the stress. If water stress develops slowly, plants may be able to speed up their life cycle, reaching maturity before the drought gets too severe. If the drought occurs rapidly, substantial damage to photosynthetic machinery may occur through oxidative stress (Ort 2001).

Water from rain or irrigation can be lost through crop transpiration, weed transpiration, soil evaporation, deep drainage, runoff, subsurface flow, or can be stored in the soil (Turner 2004). Changes in management may target these loss factors. These management strategies include reducing weeds, increasing planting density to reduce soil evaporation (Turner 2004), or decreasing soil water runoff by using drip irrigation or deficit irrigation in irrigated systems (Costa, Ortuño, and Chaves 2007). The "deficit irrigation" approach strategically provides plants with levels of water below what is considered optimal, taking into consideration stage of development and the timing of a plant's response to water stress. Such strategies are highly crop-specific, and a strong understanding of a given crop's dynamics is essential to their successful application (Costa, Ortuño, and Chaves 2007).

A complementary approach to irrigation management is to choose or develop drought-tolerant crops that have high water-use efficiency. The Fertile Crescent region was the birth-place of domesticated wheat thousands of years ago. Today, locally evolved crop varieties, or landraces, are generally cultivated in areas with high elevation and environmental stress, where dry farming is performed. They tend to be well adapted to these environmental conditions since they are exposed to many years of selection in the specific area and are therefore more likely to survive the harsh climatic conditions during seasons of extreme variability. These plants can serve as genetic stock for future crop breeding. Key traits leading to high water-use efficiency include: retaining water in the plant, rather than allowing it to evaporate at the soil surface; gaining more carbon per unit of water transpired by the crop; and storing a greater fraction of biomass in the plant component that will be harvested (Condon et al. 2004). These traits are interdependent; while one trait might be key in a given environment, it may be less important in another (Condon et al. 2004).

Effects from Changes in CO_2 Levels

The increases in atmospheric CO_2 concentrations that are largely responsible for changes in temperature and precipitation are also expected to have direct effects on plant growth. Plants must make a tradeoff when keeping their stomata open to allow CO_2 into their leaves for photosynthesis because this also causes the loss of water through transpiration. If atmospheric CO_2 concentrations increase, the pressure of this tradeoff is reduced—more growth can occur with less water loss. This "CO_2 fertilization effect" is particularly important for the group of plants that use this pathway for photosynthesis. (These are called "C3 plants," after the number of carbon atoms in the compounds involved at a key point of photosynthesis). Commonly produced crops in Jordan and Lebanon such as vegetables, fruit trees, and wheat and barley all use this pathway, while sugarcane, sorghum, maize, and some

(box continues on next page)

Box 2.1 Biophysical Impacts of Climate Change on Agricultural Systems *(continued)*

millets use a different ("C4") pathway and are therefore less sensitive to CO_2 fertilization. Jablonski, Wang, and Curtis (2002) performed a meta-analysis of studies of CO_2 fertilization effects, and found that overall, plants produced 19 percent more flowers, 18 percent more fruits, and 25 percent greater seed mass under elevated CO_2 levels.

While the CO_2 fertilization effect could potentially be beneficial, there are still key questions. These include whether it will be sufficient to offset any negative effects on yield because of temperature and water stress, how much it will be limited by other constraints such as nutrient availability (Oren et al. 2001), and whether it will favor crops over weeds (Fuhrer 2003). Research is just beginning to answer these questions and examine these complex interactions. Therefore, it will be some time before the positive or negative effects from increased CO_2 concentrations are understood with high levels of certainty.

Increased temperatures will likely result in lower carbon levels in the soil. This is a result of greater evaporation and transpiration by plants, which reduces soil moisture regardless of what the precipitation status is. Climate change will have an impact not only on soil water availability, but also organic matter cycling rates (Falloon et al. 2007) and salinity. Soil carbon levels in the Arab region are already low, and further losses because of decreased inputs or increased decomposition rates could further reduce soil water holding capacity. Increases in evaporation may increase salt accumulation in soils (Khresat 2010), which is detrimental for soil structure and plant growth. Salinity stress is expected to occur in the water systems that feed the Jordan River Valley and Bekaa Valley because with less precipitation, salt concentrations will increase (Ministry of Environment (Lebanon) 2011; Taimeh 2010).

Pest and Pathogen Management

The impact of plant pathogens or pests on crops depends on three factors: the pathogen and its characteristics, such as how virulent it is; the crop and its susceptibility or health; and the environment and whether it benefits the crop or the pest. Changes to any of these three factors can have an impact on disease severity and its net effects. As climate changes, the types and numbers of pests and diseases prevalent in a given area will change. Effects on insect pests are as complicated as the effects on their host plants, and it can be hard to predict their net impact on crops (Fuhrer 2003). Insects, being cold-blooded creatures, are often heavily influenced by temperature (Abdel-Wali 2010). Increases in temperatures may increase the number of insect generations possible each year, both because of the length of the possible growing season and the insect's accelerated development (Harvell et al. 2002). For example, a 2°C increase could result in one to five more life cycles per season (Abdel-Wali 2010). Extreme events, such as the predicted increases in droughts and floods from climate change, can act as triggers for insect outbreaks (Fuhrer 2003).

Precipitation and moisture levels are also important for the occurrence of many plant diseases. For example, leaf wetness duration is a key factor for the occurrence and spread of many leaf diseases (Juroszek and Tiedemann 2011). The germination of fungal spores and their successful infection of the plant often requires close to 100 percent relative humidity, which usually occurs during night-time dew. In addition, fungicides are often less effective

(box continues on next page)

Box 2.1 Biophysical Impacts of Climate Change on Agricultural Systems *(continued)*

under high rainfalls (Juroszek and Tiedeman 2011). This reiterates the fact that most plants require a moderate level of moisture, with too much or too little both being damaging.

An additional challenge to controlling plant disease and pests is that the efficacy of herbicides, insecticides, or fungicides may be altered with increasing temperature and CO_2 levels. For pesticides that have a gaseous form, higher temperatures could increase their volatilization, resulting in greater transport outside the targeted area and a need for greater application rates to produce the same effect. In addition, decomposition rates in soil may be changed (increased under higher temperatures, decreased under lower moisture), which would affect the persistence of these chemicals as well. If degradation rates increase, more frequent pesticide applications could be required to achieve the same effect (Bloomfield et al. 2006). Changes in characteristics of an herbicide's target, such as an increase in leaf thickness in weeds due to elevated CO_2 levels, could also reduce the herbicide's effectiveness (Juroszek and Tiedeman 2011).

Effects on Livestock

Animals are at risk from climate change in the Arab region. This is for two reasons: first, through direct physiological impacts due to high temperatures or dry conditions, and second, through the indirect effects of climate change on their food and water supplies (Easterling and Apps 2005). For example, it is known that increases in temperature beyond optimal levels leads to decreased growth rates, feed efficiency, eggshell quality, and the overall survival of poultry (Teeter and Belay 1996). Temperature stresses on dairy animals can reduce dry matter intake, leading to weight loss and increased water intake. This leads to less meat and decreased milk production (Farajalla 2010). Wolfenson, Roth, and Meidan (2000) estimate that heat stress causes economic losses in about 60 percent of dairy farms around the world. Nardone et al. (2010) show that the mean adult weight of sheep is 13.5 percent lower in Asian breeds as compared to European breeds, while weights for African breeds are 40.6 percent lower. For goats, Asian breeds are 14.4 percent lighter, and African breeds 31.7 percent lighter. While there are many factors that would affect these trends, Nardone et al. (2010) predict it is because of increasing temperatures.

Animal production can be highly water-intensive (Chapagain and Hoekstra 2003). It is estimated that the water it takes to produce 30 grams of animal protein (the daily requirements for humans) is 3.7 tons for beef, 1.9 tons for sheep, and 0.7–1.9 tons for milk (these are based on values for industrial as well as grazing production systems) (Nardone et al. 2010). Rainfed grazing systems have a much lower water impact, but are also more sensitive to water shortages. Animals need to drink more water under heat-stressed conditions. This increased water intake can have negative effects in and of itself, if the water is high in contaminants such as heavy metals, is at an unoptimal pH level, or contains excess nutrients (Nardone et al. 2010).

Fortunately, sheep and goats, key animals for both Jordan and Lebanon, are relatively heat-resistant compared to other livestock. However, at extreme or prolonged high temperatures they still experience heat stress, which reduce their milk yields (Nardone et al. 2010). Improving local breeds through selection and breeding is one potential approach to addressing this challenge (Al-Jaloudy 2006).

Notes

1. http://weather.noaa.gov/cgi-bin/nsd_country_lookup.pl.
2. http://www.ncdc.noaa.gov/cgi-bin/res40.pl.
3. http://eca.knmi.nl/dailydata/index.php.
4. http://badc.nerc.ac.uk/view/badc.nerc.ac.uk__ATOM__dataent_1256223773328276.
5. http://cerawww.dkrz.de/CERA/index.html.
6. http://www.cccsn.ec.gc.ca/?page=main&lang=en.
7. Tropical Rainfall Measuring Mission, NASA.
8. http://www.cccsn.ca/index-e.html.
9. Note that beyond the 2020s projected changes are generally greater than the combined HadCM3-SDSM downscaling bias, estimated to be a decrease of 19 percent at Amman and a 15 percent increase at Kfardane. In other words, precipitation changes given for 2020s cannot always be distinguished from natural variability and/or downscaling errors.
10. See the EU FP6 MEditerranean Development of Innovative Technologies for integrAted waTer management (MEDITATE) programme: http://www.meditate.hacettepe.edu.tr/index1/index.htm.

References

Abdel-Wali, M. 2010. "Assessing Climatic Changes in the Biotic Environment of the Agricultural System." World Bank Workshop presentation, Amman, Jordan, October.

Abdulla, F., T. Eshtawi, and H. Assaf. 2009. "Assessment of the Impact of Potential Climate Change on the Water Balance of a Semi-Arid Watershed." *Water Resources Management* 23: 2051–68.

Abu-Taleb, M. F. 2000. "Impacts of Global Climate Change Scenarios on Water Supply and Demand in Jordan." *Water International* 25: 457–63.

Alcamo, J., M. Flörke, and M. Märker. 2007. "Future Long-Term Changes in Global Water Resources Driven by Socio-Economic and Climatic Changes." *Hydrological Sciences Journal* 52: 247–75.

Al-Jaloudy, M. A. 2006. "Country Pasture/Forage Resource Profiles—Jordan." Food and Agriculture Organization of the United Nation. http://www.fao.org/ag/agp/agpc/doc/Counprof/PDF%20files/Jordan.pdf.

Al-Weshah, R. A. 2000. "The Water Balance of the Dead Sea: An Integrated Approach." *Hydrological Processes* 14: 145–54.

Arnell, N. W. 2004. "Climate Change and Global Water Resources: SRES Emissions and Socio-Economic Scenarios." *Global Environmental Change* 14: 31–52.

Baettig, M. B., M. Wild, and D. M. Imboden. 2007. "A Climate Change Index: Where Climate Change May Be Most Prominent in the 21st Century." *Geophysical Research Letters*, 34: L01705.

Bengtsson, L., K. I. Hodges, and E. Roeckner. 2006. "Storm Tracks and Climate Change." *Journal of Climate* 19: 3518–43.

Beniston, M. 2003. "Climatic Change in Mountain Regions: A Review of Possible Impacts." *Climatic Change* 59: 5–31.

Black, E. 2009. "The Impact of Climate Change on Daily Precipitation Statistics in Jordan and Israel." *Atmospheric Science Letters* 10: 192–200.

Bloomfield, J. P., R. J. Williams, D. C. Gooddy, J. N. Cape, and P. Guha. 2006. "Impacts of Climate Change on the Fate and Behaviour of Pesticides in Surface and Groundwater—A UK Perspective." *Science of the Total Environment* 369: 163–77.

Bou-Zeid, E., and M. El-Fadel. 2002. "Climate Change and Water Resources in Lebanon and the Middle East." *Journal of Water Resources Planning and Management-ASCE* 128: 343–55.

Bozkurt, D., and O. L. Sen. 2009. "Precipitation in the Anatolian Peninsula: Sensitivity to Increased SSTs in the Surrounding Seas." *Climate Dynamics* 36: 711–26. doi:10.1007/s00382-009-0651-3, online.

Bozkurt, D., O. L. Sen, U. U. Turuncoglu, B. Onol, T. Kindap, H. N. Dalfes, and M. Karaca. 2010. "Impacts of Climate Change on Hydrometeorology of the Euphrates and Tigris Basins." Poster presented at the European Geophysical Union, Vienna, Austria.

Chapagain, A. K., and A. Y. Hoekstra. 2003. "Section 3. Virtual Water Trade: A Quantification of Virtual Water Flows between Nations in Relation to International Trade of Livestock and Livestock Products." In *Virtual Water Trade—Proceedings of the International Expert Meeting on Virtual Water Trade*, edited by A. Y. Hoekstra, 49–76. Delft, the Netherlands: UNESCO-IHE Institute for Water Education.

Chaves, M. M., and M. M. Oliveira. 2004. "Mechanisms Underlying Plant Resilience to Water Deficits: Prospects for Water-Saving Agriculture." *Journal of Experimental Botany* 55 (407): 2365–84.

Christensen, J. H., B. Hewitson, A. Busuioc, A. Chen, X. Gao, I. Held, R. Jones, R. K. Kolli, W. -T. Kwon, R. Laprise, V. Magaña Rueda, L. Mearns, C. G. Menéndez, J. Räisänen, A. Rinke, A. Sarr, and P. Whetton. 2007. "Regional Climate Projections." In *Climate Change 2007: The Physical Science Basis. Contribution of Working Group I to the Fourth Assessment Report of the Intergovernmental Panel on Climate Change*, edited by S. Solomon, D. Qin, M. Manning, Z. Chen, M. Marquis, K. B. Averyt, M. Tignor, and H. L. Miller. Cambridge, UK and New York, NY: Cambridge University Press.

Condon, A. G., R. A. Richards, G. J. Rebetzke, and G. D. Farquhar. 2004. "Breeding for High Water-Use Efficiency." *Journal of Experimental Botany* 407: 2447–60.

Costa, J. M., M. F. Ortuño, and M. M. Chaves. 2007. "Deficit Irrigation as a Strategy to Save Water: Physiology and Potential Application to Horticulture." *Journal of Integrative Plant Biology* 49 (10): 1421–34.

Cullen, H. M., A. Kaplan, P. A. Arkin, and P. B. Demenocal. 2002. "Impact of the North Atlantic Oscillation on Middle Eastern Climate and Streamflow." *Climatic Change* 55: 315–38.

Cullen, H. M., and P. B. deMenocol. 2000. "North Atlantic Influence on Tigris-Eurphrates Streamflow." *International Journal of Climatology* 20: 853–63.

de Wit, M., and J. Stankiewicz. 2006. "Changes in Surface Water Supply across Africa with Predicted Climate Change." *Science* 311: 1917–21.

Easterling, W., and M. Apps. 2005. "Assessing the Consequences of Climate Change for Food and Forest Resources: A View from the IPCC." *Climatic Change* 70: 165–89.

Eshel, G., and B. F. Farrell. 2000. "Mechanisms of Eastern Mediterranean Rainfall Variability." *Journal of the Atmospheric Sciences* 57: 3219–32.

Evans, J. P. 2009. "21st Century Climate Change in the Middle East." *Climatic Change* 92: 417–32.

Evans, J. P., and R. Geerken. 2004. "Discrimination Between Climate and Human Induced Dryland Degradation." *Journal of Arid Environments* 57: 535–54.

Evans, J. P., R. B. Smith, and R. J. Oglesby. 2004. "Middle East Climate Simulation and Dominant Precipitation Processes." *International Journal of Climatology* 24: 1671–94.

Falloon, P., C. D. Jones, C. E. Cerri, R. Al-Adamat, P. Kamoni, T. Bhattacharyya, M. Easter, K. Paustian, K. Killian, K. Coleman, and E. Milne. 2007. "Climate Change and Its Impact on Soil and Vegetation Carbon Storage in Kenya, Jordan, India and Brazil." *Agriculture, Ecosystems, and Environment* 122: 114–24.

Farajalla, N. 2010. "Climate Change and Its Impact on Agriculture." World Bank Reducing Vulnerability to Climate Change in Agricultural Systems workshop presentation, Lebanon, October.

Fleury, P., M. Bakalowicz, and G. de Marsily. 2007. "Submarine Springs and Coastal Karst Aquifers: A Review." *Journal of Hydrology* 339: 79–92.

Freiwan, M., and M. Kadioğlu. 2008a. "Climate Variability in Jordan." *International Journal of Climatology* 28: 69–89.

———. 2008b. "Spatial and Temporal Analysis of Climatological Data in Jordan." *International Journal of Climatology* 28: 521–35.

Fuhrer, J. 2003. "Agroecosystem Responses to Combinations of Elevated CO_2, Ozone, and Global Climate Change." *Agriculture, Ecosystems and Environment* 97: 1–20.

Gao, X., and F. Girogi. 2008. "Increased Aridity in the Mediterranean Region Under Greenhouse Gas Forcing Estimated from High Resolution Simulations with a Regional Climate Model." *Global Planetary Change* 62: 195–209.

Giorgi, F., and P. Lionello. 2008. "Climate Change Projections for the Mediterranean Region." *Global Planetary Change* 64: 90–104.

Giorgi, F., and X. Bi. 2005. "Updated Regional Precipitation and Temperature Changes for the 21st Century from Ensembles of Recent AOGCM Simulations." *Geophysical Research Letters* 32: L21715.

Goldreich, Y. 1994. "The Spatial-Distribution of Annual Rainfall in Israel—A Review." *Theoretical and Applied Climatology* 50: 45–59.

Greenbaum, N., U. Schwartz, and N. Bergman. 2010. "Extreme Floods and Short-Term Hydroclimatological Fluctuations in the Hyper-Arid Dead Sea Region, Israel." *Global and Planetary Change* 70: 125–37.

Gregory, P. J., S. N. Johnson, A. C. Newton, and J. S. I. Ingram. 2009. "Integrating Pests and Pathogens into the Climate Change/Food Security Debate." *Journal of Experimental Botany* 60 (10): 2827–38.

Halwani, J. 2009. "Climate Change and Water Resources in Lebanon." *IOP Conference Series: Earth and Environmental Science* 6: 292011.

Harvell, C. D., C. E. Mitchell, J. R. Ward, S. Altizer, A. P. Dobson, R. S. Ostfeld, and M. D. Samuel. 2002. "Climate Warming and Disease Risk for Terrestrial and Marine Biota." *Science* 296: 2158–62.

Haylock, M. R., N. Hofstra, A. M. G. Klein Tank, E. J. Klok, P. D. Jones, and M. New. 2008. "A European Daily High-Resolution Gridded Data Set of Surface Temperature and Precipitation for 1950–2006." *Journal of Geophysical Research* 113: D20119.

Hreiche, A., W. Najem, and C. Bocquillon. 2007. "Hydrological Impact Simulations of Climate Change on Lebanese Coastal Rivers." *Hydrological Sciences Journal* 52: 1119–33.

Huffman, G. J., R. F. Adler, D. T. Bolvin, G. Gu, E. J. Nelkin, K. P. Bowman, Y. Hong, E. F. Stocker, and D. B. Wolff. 2007. "The TRMM Multi-Satellite Precipitation Analysis: Quasi-Global, Multi-Year, Combined-Sensor Precipitation Estimates at Fine Scale." *Journal of Hydrometeorology* 8: 38–55.

Jablonski, L. M., X. Wang, and P. S. Curtis. 2002. "Plant Reproduction Under Elevated CO_2 Conditions: A Meta-Analysis of Reports on 79 Crop and Wild Species." *New Phytologist* 156: 9–26.

Juroszek, P., and A. von Tiedemann. 2011. "Potential Strategies and Future Requirements for Plant Disease Management Under a Changing Climate." *Plant Pathology* 60: 100–12.

Khair, K., N. Aker, F. Haddad, M. Jurdi, and A. Hachach. 1994. "The Environmental Impacts of Humans on Groundwater in Lebanon." *Water, Air and Soil Pollution* 78: 37–49.

Khresat, S. A. 2010. "Assessing Climatic Changes in the Biotic Environment of the Agricultural System." World Bank Reducing Vulnerability to Climate Change in Agricultural Systems workshop presentation, Amman, Jordan, October.

Kim, D.-W., and H. R. Byun. 2009. "Future Pattern of Asian Drought Under Global Warming Scenario." *Theoretical and Applied Climatology* 98: 137–50.

Krichak, S. O., P. Alpert, and P. Kunin. 2010. "Numerical Simulation of Seasonal Distribution of Precipitation Over the Eastern Mediterranean with a RCM." *Climate Dynamics* 34: 47–59.

Lautze, J., and P. Kirshen. 2009. "Water Allocation, Climate Change, and Sustainable Water Use in Israel/Palestine: The Palestinian Position." *Water International* 34: 189–203.

Lionello, P., and F. Giorgi. 2007. "Winter Precipitation and Cyclones in the Mediterranean Region: Future Climate Scenarios in a Regional Simulation." *Advances in Geosciences* 12: 153–58.

Mahajan, S., and N. Tuteja. 2005. "Cold, Salinity and Drought Stress: An Overview." *Archives of Biochemistry and Biophysics* 444: 139–58.

MedCLIVAR. 2004. "White Paper on Mediterranean Climate Variability and Predictability." International CLIVAR Project Office. http://web.lmd.jussieu.fr/~li/gicc_medwater/bibliographie/MedCLIVAR_WP.pdf.

Meslmani, Y. 2008. "Vulnerability Assessment and Possible Adaptation Measures of Water Resources—Water Sector Modelling." Enabling activities for preparation of Syria's initial national communication to the UNFCCC, UNDP, 18pp.

Milly, P. C. D., K. A. Dunne, and A. V. Vecchia. 2005. "Global Pattern of Trends in Streamflow and Water Availability in a Changing Climate." *Nature* 438: 347–50.

Ministry of Environment (Lebanon). 2011. "Second National Communication to the UNFCCC." Beirut, February. maindb.unfccc.int/library/view_pdf.pl?url=http://unfccc.int/resource/docs/natc/lbnnc2.pdf.

Ministry of Environment, Jordan. 2009. "Jordan's Second National Communication to the United Nations Framework Convention on Climate Change (UNFCCC)." Amman, Jordan, 166pp.

Mitchell, T. D., M. Hulme, and M. New. 2002. "Climate Data for Political Areas." *Area* 34: 109–12.

Mizyed, N. 2009. "Impacts of Climate Change on Water Resources Availability and Agricultural Water Demand in the West Bank." *Water Resources Management* 23: 2015–29.

Moretti, C. L., L. M. Mattos, A. G. Calbo, and S. A. Sargent. 2009. "Climate Changes and Potential Impacts on Postharvest Quality of Fruit and Vegetable Crops: A Review." *Food Research International* 43: 1824–32.

Nardone, A., B. Ronchi, N. Lacetera, M. S. Ranieri, and U. Bernabucci. 2010. "Effects of Climate Changes on Animal Production and Sustainability of Livestock Systems." *Livestock Science* 130: 57–69.

Nasrallah, H. A., and R. C. Balling. 1993. "Spatial and Temporal Analysis of Middle-Eastern Temperature Changes." *Climatic Change* 25: 153–61.

Nohara, D., A. Kitoh, M. Hosaka, and T. Oki. 2006. "Impact of Climate Change on River Discharge Projected by Multimodel Ensemble." *Journal of Hydrometeorology* 7: 1076–89.

Önol, B., and F. H. M. Semazzi. 2009. "Regionalization of Climate Change Simulations Over the Eastern Mediterranean." *Journal of Climate* 22: 1944–61.

Oren, R., D. S. Ellsworth, K. H. Johnsen, N. Phillips, B. E. Ewers, C. Maier, K. V. R. Schäfer, H. McCarthy, G. Hendrey, S. G. McNulty, and G. G. Katul. 2001. "Soil Fertility Limits Carbon Sequestration by Forest Ecosystems in a CO_2-Enriched Atmosphere." *Nature* 411: 469–72.

Oroud, I. M. 2008. "The Impacts of Climate Change on Water Resources in Jordan." In *Climatic Changes and Water Resources in the Middle East and North Africa*, edited by F. Zereini and H. Hotzl. New York: Springer.

Ort, D. R. 2001. "When There Is Too Much Light." *Plant Physiology* 125: 29–32.

Pal, J. S., F. Giorgi, X. Bi, N. Elguindi, F. Solmon, X. Gao, S. A. Rauscher, R. Francisco, A. Zakey, J. Winter, M. Ashfaq, F. S. Syed, J. L. Bell, N. S. Diffenbaugh, J. Karmacharya, A. Konaré, D. Martinez, R. P. Da Rocha, L. C. Sloan, and A. L. Steiner. 2007. "RegCM3 and RegCNET: Regional Climate Modeling for the Developing World." *Bulletin of the American Meteorological Society* 88: 1395–409.

Pepin, N., and M. Losleben. 2002. "Climate Change in the Colorado Rocky Mountains: Free Air Versus Surface Temperature Trends." *International Journal of Climatology* 22: 311–29.

Ragab, R., and C. Prudhomme. 2002. "Climate Change and Water Resources Management in Arid and Semi-Arid Regions: Prospective and Challenges for the 21st Century." *Biosystems Engineering* 81: 3–34.

Raible, C. C., B. Ziv, H. Saaroni, and M. Wild. 2010. "Winter Synoptic-Scale Variability Over the Mediterranean Basin Under Future Climate Conditions as Simulated by the ECHAM5." *Climate Dynamics* 35: 473–88.

Reddy, A. R., K. V. Chaitanya, and M. Vivekanandan. 2004. "Drought-Induced Responses of Photosynthesis and Antioxidant Metabolism in Higher Plants." *Journal of Plant Physiology* 161: 1189–202.

Saadeh, M. 2008. "Seawater Intrusion in Greater Beirut, Lebanon." In *Climatic Changes and Water Resources in the Middle East and North Africa*, edited by F. Zereini and H. Hotzl. New York: Springer.

Samuels, R., A. Rimmer, and A. Alpert. 2009. "Effect of Extreme Rainfall Events on the Water Resources of the Jordan River." *Journal of Hydrology* 375: 513–23.

Samuels, R., A. Rimmer, A. Hartmann, S. Krichak, and P. Alpert. 2010. "Climate Change Impacts on Jordan River Flow: Downscaling Application from a Regional Climate Model." *Journal of Hydrometeorology* 11 (4): 860–79.

Scholze, M., W. Knorr, N. W. Arnell, and I. C. Prentice. 2006. "A Climate Change Risk Analysis for World Ecosystems." *Proceedings of the National Academy of Science* 103: 13116–20.

Sensoy, S., T. C. Peterson, L. V. Alexander, and Z. Xuebin. 2007. "Enhancing Middle East Climate Change Monitoring and Indexes." *Bulletin of the American Meteorological Society* 88: 1249–54.

Shaban, A. 2009. "Indicators and Aspects of Hydrological Drought in Lebanon." *Water Resources Management* 23: 1875–91.

Shahin, M. 2007. *Water Resources and Hydrometeorology of the Arab Region.* Water Science and Technology Library, Volume 59, 586pp. The Netherlands: Springer.

Smadi, M. M., and A. Zghoul. 2006. "A Sudden Change in Rainfall Characteristics in Amman, Jordan During the Mid 1950s." *American Journal of Environmental Sciences* 2: 84–91.

Solomon, S., D. Qin, M. Manning, R. B. Alley, T. Berntsen, N. L. Bindoff, Z. Chen, A. Chidthaisong, J. M. Gregory, G. C. Hegerl, M. Heimann, B. Hewitson, B. J. Hoskins, F. Joos, J. Jouzel, V. Kattsov, U. Lohmann, T. Matsuno, M. Molina, N. Nicholls, J. Overpeck, G. Raga, V. Ramaswamy, J. Ren, M. Rusticucci, R. Somerville, T. F. Stocker, P. Whetton, R. A. Wood, and D. Wratt. 2007. "Technical Summary." In *Climate Change 2007: The Physical Science Basis. Contribution of Working Group I to the Fourth Assessment Report of the Intergovernmental Panel on Climate Change*, edited by S. Solomon, D. Qin, M. Manning, Z. Chen, M. Marquis, K. B. Averyt, M. Tignor, and H. L. Miller. Cambridge, United Kingdom and New York, NY: Cambridge University Press.

Solomon, S., D. Qin, M. Manning, Z. Chen, M. Marquis, K. B. Averyt, M. Tignor and H. L. Miller, eds. 2007. *Contribution of Working Group I to the Fourth Assessment Report of the Intergovernmental Panel on Climate Change, 2007.* Cambridge, UK and New York, NY: Cambridge University Press.

Taimeh, A. 2010. "*Climatic Change in Jordan and Its Impacts on Land Resources.*" World Bank Reducing Vulnerability to Climate Change in Agricultural Systems workshop presentation, Amman, Jordan, October.

Tebaldi, C., K. Hayhoe, J. M. Arblaster, and G. A. Meehl. 2006. "Going to Extremes: An Intercomparison of Model-Simulated Historical and Future Change in Extreme Events." *Climatic Change* 79: 185–211.

Teeter, R. G., and T. Belay. 1996. "Broiler Management During Acute Heat Stress." *Animal Feed Science Technology* 58: 127–42.

Turner, N. C. 2004. "Agronomic Options for Improving Rainfall-Use Efficiency of Crops in Dryland Farming Systems." *Journal of Experimental Botany* 55 (407): 2413–25.

Wilby, R. L. 2010. "Climate Change Projections and Downscaling for Jordan, Lebanon and Syria: Draft Synthesis Report." World Bank, MENA Region, September.

Wilby, R. L., C. W. Dawson, and E. M. Barrow. 2002. "SDSM—A Decision Support Tool for the Assessment of Regional Climate Change Impacts." *Environmental and Modelling Software* 17: 145–57.

Wilby, R. L., and H. J. Fowler. 2010. "Regional Climate Downscaling." In *Climate Change and Water Resource Modelling*, edited by F. Fung and A. Lopez. Oxford: Blackwell Publishing.

Wilby, R. L., J. Troni, Y. Biot, L. Tedd, B. C. Hewitson, D. M. Smith, and R. T. Sutton. 2009. "A Review of Climate Risk Information for Adaptation and Development Planning." *International Journal of Climatology* 29: 1193–215.

Wolfenson, D., Z. Roth, and R. Meidan. 2000. "Impaired Reproduction in Heat-Stressed Cattle: Basic and Applied Aspects." *Animal Reproduction Science* 60–61: 535–47.

Xoplaki, E., Gonzalez-Rouco, J. F., Luterbacher, J. and Wanner, H. 2004. Wet season Mediterranean precipitation variability: influence of large-scale dynamics and trends. Climate Dynamics, 23, 63–78

Zereini, F., and H. Hötzl, eds. 2008. *Climatic Changes and Water Resources in the Middle East and North Africa*, 552pp. Berlin: Springer.

Zhang, X., E. Aguilar, S. Sensoy, H. Melkonyan, U. Tagiyeva, N. Ahmed, N. Kutaladze, F. Rahimzadeh, A. Taghipour, T. H. Hantosh, P. Alpert, M. Semawi, M. K. Ali, M. H. S. Al-Shabibi, Z. Al-Oulan, T. Zatari, I. Al Dean Khelet, S. Hamoud, R. Sagir, M. Demircan, M. Eken, M. Adiguzel, L. Alexander, T. C. Peterson, and T. Wallis. 2005. "Trends in Middle East Climate Extreme Indices from 1950 to 2003." *Journal of Geophysical Research Atmospheres* 110: D22104.

Ziv, B., U. Dayan, Y. Kushnir, C. Roth, and Y. Enzel. 2006. "Regional and Global Atmospheric Patterns Governing Rainfall in the Southern Levant." *International Journal of Climatology* 26 (1): 55–73.

CHAPTER 3

Priority Setting for Building Agricultural Resilience

Photograph by Dorte Verner

Methodology

Climate change poses many challenges for developing countries, but at the root of these challenges lies the vulnerability of human populations. The IPCC (2001) defines **vulnerability** as "a function of the **sensitivity** of a system to changes in climate, **adaptive capacity**, and the degree of **exposure** of the system to climatic hazards." Ashwill, Flora, and Flora (2011) define each variable as the following: "**exposure** as the exogenous drivers of vulnerability, or climate related events and changes that humans cannot directly control (...) **Sensitivity** refers to all of the endogenous drivers of vulnerability, these include community characteristics or practices that humans can control and which contribute to vulnerability (...) **Adaptive capacity** encapsulates the community characteristics or practices that contribute to building resilience and reducing vulnerability." This vulnerability has many sources: the tropical locations of many developing nations and the prevalence of drought, flooding and other natural disasters; widespread food and health insecurity; poor infrastructure, making it difficult to provide transportation, communications and, under circumstances of crisis, emergency services. Weak and ineffective institutions, both governmental and non-governmental, are pervasive. For agricultural and rural populations, vulnerability is further exacerbated by their underlying dependence on agriculture and natural resources, the inherent sensitivity of agriculture to climatic conditions, widely degraded soils, and the prevalence of rural poverty.

Dealing with these multiple sources of vulnerability is challenging under the best of circumstances. As a result, developing informed strategies and establishing priorities among the many possible investments, interventions, policies and other adaptation mechanisms is a primary challenge confronting climate adaptation. This is especially true given the severe resource constraints faced by many developing nations (Kuch and Gigli 2007). A variety of priority-setting methods have been developed to help rank main concerns and allocate resources in many areas, including public and mental health interventions (see, for example, the review in Baltussen and Niessen 2006); public health research; vaccine development (Institute of Medicine 1986); research investments in agriculture (Alston, Norton, and Pardey 1995) and biotechnology (Falconi 1999); and various other applications, especially where the allocation of public sector budgets among competing demands is involved. This includes a recent application of a formal priority-setting approach to addressing climate change adaptations in Latin American agriculture (World Bank 2009). Priority-setting methods include a wide variety of formal approaches, with varying levels of quantitative sophistication. These include (Alston, Norton, and Pardey 1995; Falconi 1999; Hartwich 1999; World Bank 2009):

- **Congruence methods**, which rank alternative investments, choices, or themes by a single measure
- **Multi-criteria scoring methods**, which evaluate alternative choices according to multiple criteria, which are identified and weighted and then used in evaluating the initial choices
- **Capital budgeting methods** (including net present value, internal rate of return, etc.), which rank alternatives by assessing their (discounted) financial benefits and costs
- **Economic surplus analysis**, which uses applied welfare economics measures to examine the welfare gains and losses of various groups (producers, consumers, input suppliers, government) affected by an investment or intervention and then estimates the net benefits and their distribution
- **Simulation models**, which optimize single or multiple priority objectives, incorporating alternative variables, constraints and risk and uncertainty conditions, and then estimate the outcomes of alternative scenarios and the
- **Analytic hierarchy process**, a multi-objective, multi-criteria decision-making tool that uses multiple paired comparisons to rank alternative solutions to a problem, which are formulated in hierarchical terms.

For this report, it was determined to employ the priority-setting approach developed by the World Bank (2009) and applied to three countries in Latin America (Mexico, Peru and Uruguay) with regard to climate adaptation in agriculture. This approach uses science-based climate projections and a participatory multi-criteria scoring method to identify and prioritize alternative strategies for agricultural climate adaptations. Although many variations of multi-criteria scoring methods exist, the approach typically involves several steps. First, the alternative options that are to be ranked—for example, public investments, research programs, or policy options—must be identified. Second, the specific **criteria** by which the interventions will be subsequently assessed must be identified and weighted. Third, using a performance matrix, each intervention option is scored by each criterion in a "matrix ranking" process. Finally, the scores are multiplied by the criteria weights to calculate a composite weighted score that is used to prioritize each option. These priorities are then commonly incorporated into strategic plans and resource allocation processes, or used to backstop policy decisions involving the selection or ranking among competing alternatives.

The advantages and disadvantages of scoring methods vis-à-vis alternative approaches for priority setting are well established (Alston, Norton, and Pardey 1995; Falconi 1999; Manicad 1997; World Bank 2009). Advantages include the fact that the approach is relative easy to understand, to apply and is transparent; it can be used to incorporate widely different types of evaluation criteria, quantitative and qualitative; the approach lends itself to the active participatory involvement of stakeholders; and finally, it does not require advanced quantitative skills, in contrast to the economic surplus or simulation methods, for example. Additionally, as shown in the earlier World Bank study (2009), the scoring approach—when applied in a workshop-type setting—provides a constructive mechanism for stakeholders with widely diverse backgrounds, experiences and views to try to reconcile alternative opinions. The disadvantages of scoring methods are also well known. These include their simplicity and possibly inconsistent treatment of measurable quantitative criteria (which can be included in the assessment, but need not be); the difficulties presented by overlapping objectives and criteria; their limitations in discounting future benefits and costs; and the sensitivity of the results to the selection of the participants doing the scoring and their subjective beliefs and biases.

As noted in the previous application of this approach (World Bank 2009), scoring and similar approaches are sometimes criticized for their lack of explicit inclusion of cost-benefit estimates and their potential to develop priority-setting schemes. This criticism stems from these approaches not being based on comprehensive economic estimates that often constitute the standard basis for project assessments. While there is no reason a priori that multi-criteria scoring approaches cannot include explicit cost-benefit criteria, they often do not. For this report, an attempt was made to incorporate rough cost estimates in the development of some of the "profiles" of alternative response options and indeed, there is nothing necessarily inconsistent in combining both approaches in this way. However, from the beginning, in this and in many similar applications, multi-criteria priority-setting approaches are rarely planned to yield the detailed quantitative estimates of costs and benefits that one finds in most cost-benefit analyses. Rather the approach typically focuses on incorporating diverse evaluation criteria, economic and non-economic, quantitative and qualitative. Additionally, it should be noted that cost-benefit analysis is often most useful and appropriate in a deterministic setting where uncertainty is low (Environmental Assessment Institute 2006; Pizer 2005). This hardly applies in the case of climate change in agriculture. Climate change and its effects are characterized by many sources of uncertainty—from the underlying causes of climate change (natural vs. anthropocentric), to the complex interaction effects characterizing many climate change impacts, to the uncertain responses of adaptation and mitigation strategies. While cost-benefit analysis has much to offer, the high levels of uncertainty that characterize climate change adaptation at every level mean that even the most complete cost-benefit studies will yield less than definitive results. Accordingly, multi-criteria priority setting and cost-benefit analysis may best be viewed as complementary approaches. As indicated previously, the priority-setting approach used for this report can most usefully be considered a "first step" in priority setting for agricultural adaptations to climate change in the study countries.

The priority-setting approach applied for this report to study climate adaptations in the Middle East was revised slightly from the prior World Bank study (2009). That approach was, in turn, adapted from the regional research planning priority-setting methodology developed by Janssen and Kissi (1997), variants of which have been widely used in agricultural research planning and priority setting in many countries. This methodology also borrows from priority-setting approaches applied in other areas, including the "interactive bottom-up" approach applied to priority setting in agricultural biotechnology research (Commandeur 1997) and the "stepped agro-ecological" approach developed for priority setting in agroecological research (Thiombiano and Andriesse 1998). The overall approach used here is based on several steps:

- Developing regional climate projections to backstop the prioritization of adaptation response options
- Formation of a local country team to implement the priority-setting process
- Use of baseline information on climate changes and their impacts in agriculture to support the project throughout its duration, including the identification of response options, policy and institutional interventions, etc.
- Sharing and use of this information with stakeholders and participants in priority-setting workshops and ideally
- Sharing the results of the overall priority-setting process through meetings with appropriate officials and policy makers.

The activities in this study were built around a series of four steps, executed in a series of workshops and meetings, all designed to lead to the development of a draft regional action plan. These steps were sequenced as follows:

Step 1   (Workshop 1): a comprehensive review of expected climate changes over the next 30–40 years (and beyond) in each study region, their major expected implications for agricultural systems in the region, and the presentation of both at a first workshop

Step 2   (Workshop 1): identification by stakeholders and workshop participants of a set of potential response options to address the climate change challenges identified in Step 1

Step 3   (Workshop 2): development of profiles of selected response options by the country team prior to, and their presentation in, Workshop 2; identification and weighting of evaluation criteria by stakeholders, and their scoring and prioritization of the final set of response options

Step 4   (Action plan development): Using the final set of priority response options, development of a draft action plan for subsequent presentation to, and consideration by, national and regional policy makers and decision makers.

The four-step priority-setting process is built around a series of workshops and meetings with stakeholders and policy makers. The local country teams played an indispensable role planning and organizing each of the workshops and meetings, inviting presenters and participants, endeavoring to get a diverse, yet representative set of stakeholders to participate in each workshop, and preparing the response option profiles and draft Action Plan that constituted the final outcome of the study. In Jordan, leadership of the study and country team coordination was through the National Center for Agricultural Research and Extension (NCARE). In Lebanon, the Lebanese Agricultural Research Institute (LARI) undertook the central organizing role for the study. Scientists, researchers, and officials from other national government institutions, universities, and other organizations played active roles as presenters and advisers at the study workshops. Participants at the workshops were very diverse, consisting of farmers and representatives of farmer organizations, research scientists, extension specialists, university personnel, representatives from nongovernmental organization (NGOs) and local governments, journalists, and representatives from international development and research organizations. External consultants and World Bank staff also played an important support role in the workshops.

Lebanon

Agriculture in Lebanon and the Bekaa Valley

In 2008, approximately 286,000 hectares of Lebanon was classified as arable land and permanent crops (World Bank 2012). The most important agricultural region of the country, containing more than 40 percent of its cultivated land, is centered

Table 3.1 Bekaa and Other Departments (*Mohafaza*) in Lebanon

	Number of growers		Cultivated land		Irrigated cultivated land	
Mohafaza	Total	%	Area (ha.)	%	Area (ha.)	%
Bekaa	35,146	18	102,948	42	53,662	52
North Lebanon	56,538	29	63,728	26	25,489	24
Mount Lebanon	42,146	22	25,667	10	9,971	10
Nabatiye	32,495	16	26,026	10	2,144	2
South Lebanon	29,504	15	29,570	12	12,743	12
Total	**195,829**	**100**	**247,939**	**100**	**104,009**	**100**

Source: Ministry of Agriculture (Lebanon) and FAO 2000 (Agricultural Census 1998).

in the Baalbek-Hermel district in the Bekaa Governorate. This governorate covers an area of 4,429 square kilometers. At the time of the last national Agricultural Census (1998), roughly 35,000 growers were cultivating approximately 103,000 hectares in the Bekaa Valley, just over half of which was under irrigation (table 3.1) (Ministry of Environment [Lebanon]/LEDO/ ECODIT 2001).

The Bekaa Valley, one of the two focus regions of this report, forms an extension of the Jordan Rift Valley, and is located about 30 kilometers (19 miles) east of Beirut, the capital of Lebanon (see map 3.1). The Bekaa Valley is about 120 kilometers (about 75 miles) in length and has an average width of about 16 kilometers (about 10 miles). It is flanked by two mountain ranges: Mount Lebanon to the west and the Anti-Lebanon mountain range to the east. These mountain ranges feed two major rivers in the Bekaa Valley: the Orontes, which flows northeast to the Syrian border, and the Litani, which flows southwest before emptying into the Mediterranean in southern Lebanon.

Agriculture in the Bekaa Valley is severely constrained by the physical nature of the land. The mountain ranges lying to the east and the west, high population density, traditional land tenure patterns, and rapidly increasing urbanization are jointly responsible for landholdings varying between 2 and 5 hectares (Hamadeh et al. 1999), with an average at the time of the last Agricultural Census (1998) of 2.9 hectares (Ministry of Environment (Lebanon)/LEDO/ ECODIT 2001). This, however, is significantly larger than the average landholdings in Lebanon; at the time of the 1998 Agricultural Census, 75 percent of Lebanon's farmers were estimated to be cultivating less than 1 hectare; 87 percent of farmers cultivated less than two hectares (FAO 2009).

The geographic regions of the Bekaa Valley can be classified according to precipitation: **Northern Bekaa** is arid with 250–275 millimeters a year; **Central Bekaa** (the Central Valley) is semi-arid with 500–600 millimeters a year; and **Southern Bekaa** is non-arid with 700–750 millimeters a year (Amery 2002). The peak of the rainy season is between January and April, where 75 percent of rainfall occurs. Average temperatures in the Bekaa Valley range from 9°C in the winter to 27°C in the summer.

The soils, which include Alfisols, Inceptisols and Aridisols, are rich in calcium carbonates (5–30 percent) with appreciable amounts of available potassium. Active lime varies between 20 and 30 percent and the pH is generally high,

Map 3.1 Regions of Lebanon

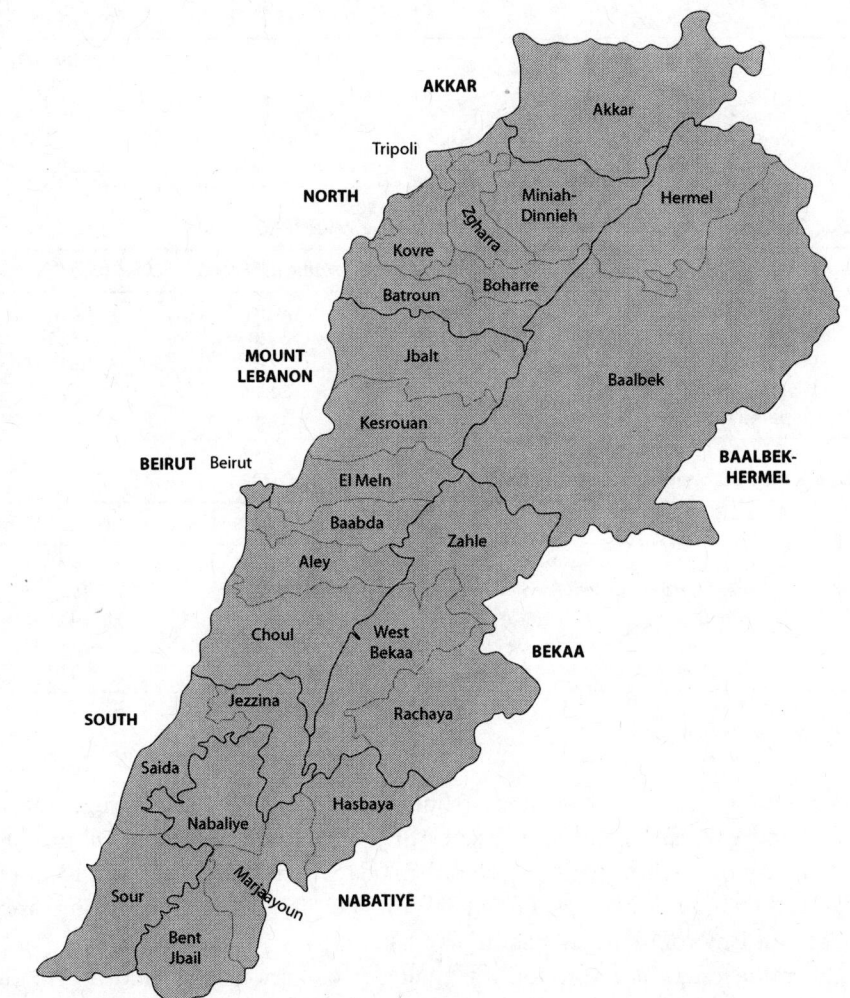

Source: World Bank data.

between 8 and 9, which reduces the availability of metal macronutrients. Phosphorus is also deficient except in fields routinely fertilized with the element, in which case a strong residual supply is found. The amounts of organic matter and of nitrogen are generally low (1 percent at depth less than 0.5 meter) and nitrogen deficiency is common. Soil depth is highly variable, affecting moisture storage capacity and causing variable maturity dates in areas of similar rainfall (Hamadeh et al. 1999).

Agriculture occupies 67.3 percent of available land in Lebanon—the highest relative proportion of agricultural land in the Arab region (World Bank 2012). Lebanon's main agricultural products are citrus fruits, grapes, tomatoes, apples, vegetables, potatoes, olives, tobacco, poultry, sheep, and goats (table 3.2). Agricultural production is concentrated in the Bekaa Valley, which accounts for

Table 3.2 Major Agricultural Production Sectors, Lebanon, 2010

Rank	Commodity	Production (US$1000)	Production (MT)
1	Indigenous chicken meat	198,091	139,069
2	Tomatoes	102,739	278,000
3	Potatoes	92,880	574,100
4	Almonds, in shell	84,103	28,500
5	Cow milk, whole, fresh	78,171	250,500

Top 5 agricultural commodity exports from Lebanon, by value (FAO 2010)

Rank	Commodity	Quantity (tons)	Value (US$1000)	Unit value (US$/ton)
1	Food Prep, nes	16,159	30,217	1,870
2	Sugar Confectionery	4,544	24,374	5,364
3	Vegetables Preserved, nes	13,509	22,658	1,677
4	Beverage Non-Alc	27,194	22,375	823
5	Prepared Nuts (exc. groundnuts)	5,657	21,677	3,832

Top 5 agricultural commodity imports to Lebanon, by value (FAO 2010)

Rank	Commodity	Quantity (tons)	Value (US$1000)
1	Cigarettes	10,193	167,901
2	Meat-Cattle boneless (beef & veal)	40,902	152,591
3	Wheat	537,692	108,886
4	Food prep	25,576	97,144
5	Sugar refined	166,118	83,571

Source: FAO 2010.
Note: nes = not elsewhere specified.

62 percent of the land dedicated to industrial crops (including sugar beets, tobacco, and vineyards) and 57 percent of the land dedicated to cereal production Ministry of Environment (Lebanon) 2008). The Akkar Valley and Koura in northern Lebanon host 40 percent of the country's olive-producing areas (Ministry of Environment (Lebanon) 2008).

High-value crops account for the majority of agricultural land-use in the Bekaa Valley, including vegetables and legumes (26 percent of total cultivated areas), fruit trees (26 percent), olive trees (3 percent), industrial crops (2.5 percent), and other (5.5 percent) (Ministry of Agriculture (Lebanon) 2007). Conditions are also favorable for the cultivation of cereals, mainly wheat and barley, which accounts for 36 percent of cultivated area in the Bekaa Valley. In 2005, 70 percent of the country's total production of cereals (roughly 400,000 tons) came from the Bekaa Valley (Ministry of Agriculture (Lebanon) 2007).

Lebanon is self-sufficient in poultry, fruit and vegetables, and produces 45, 15, and 10 percent, respectively, of its legumes, wheat, and sugar needs. Top Lebanese agricultural exports include processed food, fresh and processed fruits and vegetables, wine, and olive oil. Top agricultural imports include cigarettes, meat, wheat, processed food, sugar, and cheese. The country imports 78 percent of its dairy and meat products (FAO 2009).

The Baalbek-Hermel region is also the most important area for livestock production in the country. With a large and expanding forage area, cattle

(27,000 head), goat (272,000 head), and sheep (265,000 head) rearing is concentrated on the eastern slopes of the valley, where soil fertility is relatively low.
Main livestock products include: red meat of different varieties, poultry meat,
milk and its derivatives, eggs, honey, and fish (Ministry of Agriculture (Lebanon)
2007). As a percentage of total production, the Bekaa Valley produces an estimated 35 percent, 55 percent, and 78 percent, respectively, of the nation's total
cattle, goat, and sheep production (Ministry of Agriculture (Lebanon) 2007).

Lebanon is relatively rich in water resources with more than 2,000 rivers and
usable surface and groundwater reaching 2.2 billion cubic meters (Qumair
1998, cited by Houri and El Jeblawi 2007). In 2005, total water withdrawals
were estimated at 1,310 million cubic meters. This included 60 percent for
agricultural purposes, 29 percent for domestic use and 11 percent for industrial
use (FAO 2009). A significant amount of this water was used to irrigate agricultural land, which in 2003 totaled 105,293 hectares (Ministry of Agriculture
(Lebanon) and FAO 2000). Annual crops—principally cereals, potatoes, citrus,
and vegetables—are grown on 77 percent of the total irrigated land. Irrigation
systems and associated water use occur disproportionately on large farms.
According to the 1999 National Agricultural Census, 60 percent of cultivated
land exceeding 10 hectares was irrigated, but only 42 percent for farm holdings
between 4 and 10 hectares were irrigated.

The Bekaa Valley contains 12 of Lebanon's 30 aquifers (Mehmet and Bicak
2002). A total of 670–875 million cubic meters of water is withdrawn every
year in this region for irrigation purposes (ACS 2006). Annual fresh water withdrawals for agriculture in 2002 amounted to more than 80 percent in the Bekaa
Valley (Karam and Karaal 1999). The fields in the Bekaa Valley are heavily
irrigated with 67 percent of all crops being irrigated, compared to a national
average of 49 percent (Ministry of Agriculture (Lebanon) 2007). Over 90 percent of water for irrigation in Baalbek-Hermel is sourced from the ground.
Sprinkler irrigation is practiced in areas with potatoes and sugar beets in the
Bekaa central plain, and micro-irrigation is mostly practiced on vegetable crops,
particularly in North Bekaa (Qaa region) (FAO 2007). The sustainability of this
rate of water use is questionable, however. The Bekaa Valley was estimated to
consume 1.5 times the annual ground and surface water replenishment, leading
to declining groundwater tables (Irrigation in the Near East Region, 1996).

Agricultural activity in Lebanon is considered one of the most productive in
the Mediterranean region (Hassine and Kandil 2009). Karam and Karaal
(1999) state that this relatively high productivity is due, in great part, to the
utilization of modern irrigation technologies. It is also largely because of high
technical efficiency and Total Factor Productivity (TFP) growth, especially in
fruit, citrus, and vegetable production. For example, Lebanon's technical efficiency in fruit production is the highest among Arab countries and the second
highest in the Mediterranean region, after France (Hassine and Kandil 2009).
These high levels of technical efficiency have been attributed to trade openness
(measured by imports of agricultural equipment) and high human capital
levels (Hassine and Kandil 2009).

Approximately 8 percent of Lebanon's population—roughly 300,000 people—live under conditions of extreme poverty, meaning that they are unable to meet basic food and non-food needs (Laithy, Abu-Ismail, and Hamdan 2008). Poverty in Lebanon is mostly an urban phenomenon, with only 25 percent of people living below the national poverty line being rural (International Fund for Agricultural Development 2003). Northern and southern Lebanon are the poorest regions, with the Bekaa Valley not far behind.

Economic growth rates across the country have varied highly in the past 10–15 years. The distribution of household incomes (proxied by expenditures) is relatively unequal, although in line with income inequality in the MENA region; most inequality (92 percent) is within-governorates versus across governorates (Laithy, Abu-Ismail, and Hamdan 2008). Poverty is aggravated by a number of factors: high rates of unemployment, especially among youth and unskilled workers; low levels of educational attainment and school participation among the poor (45 percent of household heads in poverty have less that an elementary education); and low literacy among the poor. The northeastern district of Baalbek-Hermel is the poorest in the Bekaa Valley, with over 30 percent of the population living in poverty (Laithy, Abu-Ismail, and Hamdan 2008). Agriculture provides an important source of livelihood to the families tending to over 20,000 farms in the district. Conditions of poverty in Lebanon were exacerbated by the 2006 Israel–Hezbollah War, but in rural areas it is also largely because of chronic factors such as small landholdings and underperforming cereal and livestock sectors relative to more highly productive high-value crops. At the time of the 1998 Agricultural Census, two-thirds of Lebanon's farmers were estimated to be only partially employed in agriculture, with most having second jobs.

Impacts of Climate Change on Agricultural Systems in the Bekaa Valley

Lebanon has a typical Mediterranean climate, with hot and dry summers (cooling off at night), and warm and wet winters, with most rain occurring after December (Collelo 1987). Humidity is high along the coast, while the Bekaa Valley is drier and cooler than the rest of the country because it is shielded from the sea by the Lebanon Mountains. There is a wider variation in temperature both daily and annually in the valley than along the coast (Collelo 1987). Hence, the region depends substantially on irrigation to grow crops, and the long dry summers commonly cause water shortages (Sheehan 2008). The Bekaa Valley contains 46 percent of Lebanon's cultivated land (FAO 2011) up from 42 percent in 2000 (table 3.1). Pressure on the land base has led to a decline nationally in wheat production in favor of high-value crops such as vegetables. However, for the Bekaa Valley, grains, sugar beets, grapes, and livestock remain key agricultural products (Verdeil, Faour, and Velut 2007).

Temperature Effects

Predicted temperature changes in Lebanon are similar to those predicted for Jordan (FAO 2011). Farajalla estimates that the annual minimum temperature in Beirut has increased by 2.9°C over the past 125 years (Farajalla et al. 2010).

Lebanon's Second National Communication to the UNFCCC (Ministry of Environment 2011) predicts that maximum temperatures could increase by as much as 1.8°C by 2036, with minimum temperatures close behind (a 1.5°C increase). Extreme temperatures can cause significant harm to crops. In 2010, a combination of heat, drought, and fires caused wheat yields in Lebanon to decrease by 83 percent (Joumaa 2010).

Because the Bekaa Valley is currently cooler than other regions of the country, temperature increases may allow the introduction of new crops in this region. Higher temperatures in recent years have allowed citrus trees to thrive in the mountains, where they were previously not viable. At the same time, apple, cherry, peach, and grape crops in these regions have been harmed by the higher temperatures, decreasing their yields (Joumaa 2010). For wheat, the effect of increased spring temperatures reducing grain-filling time, in combination with evapotranspiration decreasing soil moisture, is predicted to decrease yields after 2050 (Ministry of Environment (Lebanon) 2011).

Sugar beets are also an important crop in the Bekaa Valley. Optimal temperatures for sugar beets rest between 17°C and 25°C, with cooler temperatures best for sugar accumulation and warmer temperatures optimal for leaf growth and photosynthesis. Fortunately, sugar beets are known to be relatively resistant to high temperatures, dry conditions, and mild salinity, to the point where a harvestable crop yield can still be obtained under conditions where other crops, such as wheat or barley, would experience total crop failure (Ober and Rajabi 2010). Still, high temperatures can cause damage such as leaf scalding and premature aging. Both heat and water stress act in concert in drought conditions, but Ober and Rajabi (2010) note that Qi and Jaggard (2006) partitioned yield losses between the two causes, and found that the impact of heat stress was of similar importance to the impact of water stress alone.

While predicted climatic changes may reduce the risk of frost for tomatoes and potatoes, the mean temperature may reach points above 30°C in the summer, which is outside of their optimal range of 10–30°C. These changes could result in earlier potato planting in winter, which could save irrigation water and increase yields (Haverkort 2008), and allow for a second crop in the autumn, between September and December. However, growing in these seasons carries its own risks, as winter cultivation could increase the risk of nematodes, aphids, late blight, brown rot, and erwinia, due to higher humidity and temperatures (Haverkort 2008; Ministry of Environment 2011).

Cherries are sensitive to temperatures above 21°C, and production in the Bekaa Valley is predicted to be highly vulnerable to climate changes. Chilling time may be barely sufficient by 2024, and may be deficient by 2100. Predicted temperature increases would increase the risk of blossom pollination failure by up to 50 percent in the Bekaa Valley. Apple trees are considered highly vulnerable to climate change in Lebanon. Similarly to cherries, chilling time is predicted to become insufficient, resulting in dysfunctional opening of tree buds (Ministry of Environment 2011). Fruit trees are also at risk of premature dropping of fruits, caused by high temperatures (Joumaa 2010). Grapes may be relatively resistant to temperature changes. Table grape vines can tolerate maximum temperatures over 40°C, although prolonged temperatures above 30°C can reduce fruit quality (Ministry of Environment 2011) and cause heat damage (Joumaa 2010). Earlier development may affect grapes negatively, increasing the risk from spring frosts, early ripening, and sunburn (Ministry of Environment 2011).

Precipitation Effects

In Lebanon, agriculture uses 60–70 percent of the country's available water (Ministry of Environment (Lebanon) 2011). Population growth and industrialization are increasing pressure on water sources; demands in the Mediterranean region have increased by 60 percent in the last 25 years. Lebanon is already experiencing substantial changes in water availability: Shaban (2009) estimates that rainfall and snow cover have decreased between 12 percent and 16 percent in the last 40 years, rivers and groundwater between 23 percent and 29 percent, springs by 43 percent, and local reservoirs by 79 percent. Importantly, the greatest decreases occur at sources where not only climate changes are occurring, but also human pressures are increasing (Shaban 2009). Bou-Zeid and El-Fadel (2002) predict a maximum decrease in available water of 15 percent, coupled with an increase in agricultural demand of 6 percent as soon as 2020 in Lebanon. Increasing future water scarcity suggests the importance of increasing water use efficiency in agriculture, including the development of new irrigation and water storage technologies and the expanded use of improved water management practices.

Wheat yields are strongly coupled to rainfall; ideally minimum annual rainfall should be above 400 millimeters. The Bekaa Valley already has relatively frequent periods of low rainfall, compared to the rest of the country, so wheat yields in this region may be particularly sensitive to climate changes. The onset of the rainy season is predicted to change, and because it determines the sowing date for wheat, is expected to result in a shorter growing season (Ministry of Environment (Lebanon) 2011).

Virtually all potato and tomato cultivation in the Bekaa Valley is irrigated. Summer cropping of potatoes is highly vulnerable to water availability, as tuber formation could be compromised if irrigation is low (Ministry of Environment (Lebanon) 2011). Potato growers in the Bekaa Valley have strong cultural traditions that dictate when irrigation is needed, which may not always be based on scientific understanding or specific soil types. Changes in precipitation trends may mean that traditional knowledge about when to irrigate is less relevant in the future, creating the need for other systems of determining when irrigation should be applied (Zeid 2005).

Cherries and apples grown in the Bekaa Valley are also predominantly irrigated. Reductions in irrigation during May-July have the potential to increase fruit drop rate and reduce overall quality (Ministry of Environment (Lebanon) 2011). There is a successful drought-resistant rootstock (*Prunus mahaleb*), that has been used to enable rainfed or low-irrigation cherry production, which could help to reduce impacts from lower irrigation water availability (Ministry of Environment (Lebanon) 2011). Although only 30 percent of apple crops use drip irrigation across Lebanon, those in the Bekaa Valley are more likely to use such systems, due to water scarcity. This is in contrast to the traditional "flood irrigation" system, where water is applied to basins dug around the trees. Although drip irrigation has been proven to be successful at reducing water inputs, many growers are hesitant to adopt it, due to either perceived abundance of water, implementation costs, the existing water allocation system, or a lack of awareness of drip irrigation (Zeid 2007). So there is considerable scope to increase water use efficiency with the expanded use of drip irrigation.

Grapes are also a key crop for the Bekaa Valley. While table grapes are mostly irrigated, industrial production systems are generally rainfed. Fortunately, grapevines are relatively resistant to warm and dry conditions. They can tolerate drought well, and persist under rainfall below 300 millimeters per year (Ministry of Environment (Lebanon) 2011). However, where growers do irrigate grapes, irrigation is not generally scientifically practiced, and water may be applied at inopportune times, resulting in increased weed competition or delayed development of sugars in fruits (Zeid 2005).

Although relatively drought-tolerant compared to other plants, the sugar beet is still susceptible to water stress. Ahmadi et al. (2011) found mean sugar root yield reductions of 35 percent and sugar yield reductions of 43 percent in a study of 76 genotypes of sugar beets subjected to drought stress in Iran. However, there is a wide range of drought stress tolerance in sugar beet cultivars, which might be exploited to develop more water stress-resistant varieties (Ahmadi et al. 2011; Ober and Rajabi 2010).

CO_2 *Effects*

Key crops in the Bekaa Valley are predominantly C3 crops, which are predicted to respond positively to increased atmospheric CO_2 concentrations. Manderscheid, Pacholski, and Weigel (2010) found that increasing CO_2 concentrations to 550 parts per million for sugar beets led to increased total biomass by 7–12 percent and increased sugar yield by 12–13 percent. However, they suggest that this response may be limited by the potential growth rate of the beet root ("sink-limited"). Under temperatures higher than 18°C, the growth rate of its storage root decreases, which Manderscheid, Pacholski, and Weigel (2010) predict had effects in their study. As discussed previously, any CO_2 fertilization impact must be considered within the context of the *net impact* of temperature, precipitation, CO_2 increases, and their combined secondary effects, making quantitative predictions challenging.

Another important crop for Lebanon is the potato. Fleisher, Timlin, and Reddy (2008) studied the interaction effects between CO_2 enrichment and water stress on potatoes, with 370 parts per million and 740 parts per million (very high) CO_2 levels, and between 10 percent and 100 percent of daily water uptake in the control plots. While CO_2 enrichment tended to enhance belowground biomass, decreased water availability decreased belowground biomass. The CO_2 enrichment effects appeared to mitigate the drought effects on belowground yields, except for the driest treatments. Interestingly, the trends in potato plant growth were clearer for the aboveground biomass, indicating that changes in how the plant partitions its biomass—above versus below ground—may be occurring.

Pests and Pathogen Management

Recent changes in pest impacts have also been observed. Choueiry and Hobaika (2010) report impacts on diverse crops, which they ascribe to climatic changes. Wheat crops in the Bekaa Valley have experienced general yellowing, root rot, stem blackening, *Fusarium sp.* infections, rust, and an unusual prevalence of insect pests such as aphids and thrips. Potatoes in the Akkar Valley to the north have been more affected by late blight in recent years due to high temperatures. Increased prevalence of thrips, which are the vector for the tomato spotted wilt virus (TSWV), has affected potatoes and tomatoes in the Akkar Valley. Apples have suffered from increased populations of red mites, which have been able to complete more generations than usual in a given year because of climatic changes (Choueiry and Hobaika 2010). While any individual observation may not necessarily be conclusively linked to long-term climate change, taken together, they indicate the types of changes that are occurring today, and may occur in the future.

As mentioned above, intensively farmed fruits, vegetables and other high-value crops are of dominant importance in Lebanese agriculture, and thus improved management holds significant potential for gains to farmers and the sector as a whole as the climate changes. For example, fruit crops and vegetables are frequently intensively sprayed with various pesticides, often under a "better safe than sorry" mentality. There is limited monitoring of pests and relevant environmental conditions to inform pesticide applications, often resulting in applications at the maximum levels (Zeid 2007). Powdery mildew is a common problem for apples and grapes, however, as it thrives under higher humidity, it is somewhat less of an issue in the dry Bekaa Valley, and might be inhibited under drier conditions (Zeid 2005). Similarly, the late blight, which can be devastating to potatoes and tomatoes, is a fungus that is typically less of an issue under drier conditions (Zeid 2005). However, increased spring temperatures are predicted to increase the rate of cherry fly infestation (Ministry of Environment (Lebanon) 2011).

Effects on Livestock

Meat and milk from goats, sheep, and cows, are prevalent agricultural products in Lebanon, but are secondary to other forms of agricultural production (Asmar 2011). Poultry and egg production are also common. Goats and sheep are particularly concentrated in the Bekaa Valley. Effects of overgrazing and land fragmentation due to urban sprawl have decreased herd numbers. In regions of Mount Lebanon, this decrease in grazing has subsequently led to increased biomass growth, and, with it, increased intensity and frequency of forest fires (Asmar 2011). While there are diverse breeds of varied characteristics, which could provide genetic stock for breeds resistant to climate changes, the wild goat of the Lebanese mountains has disappeared and may be extinct (Asmar 2011). Regarding milk production, relatively few farmers have cooling facilities, meaning that any climatic changes that further promote the growth of bacteria or other contaminants in milk could have a negative impact on milk quality and value (Asmar 2011). The area planted for forage crops in Lebanon, mainly concentrated in the Bekaa Valley and the Akkar Plains, has increased over the past decade, but most feed is imported. Any impacts on these crops—mostly barley and vetch—could pass impacts on to the livestock and farmers that depend on them.

Project Description and Results: Bekaa Valley, Lebanon

As discussed above, the priority-setting project in Lebanon centered on the completion of four steps, involving two workshops and the development of a draft action plan to address climate adaptation in agriculture in the Bekaa Valley. Each step is briefly discussed next in this chapter.

Step 1, Workshop 1: Review of Climate Change and Impacts on Agriculture in the Bekaa Valley

The project workshops in Lebanon and Jordan followed the same overall structure outlined above in the methodology. The first workshop in Lebanon was held in Baalbeck in the Bekaa Valley over two days in October 2010, and consisted of two parts. The first part (Step 1) consisted of several presentations by national scientists, researchers from the Lebanese Agricultural Research Institute (LARI) and other experts reviewing the scientific literature and empirical evidence on (1) climate changes in the Middle East, in general, and in Lebanon specifically, and (2) the impacts of climate change on agriculture, on resources—including, water, soil, and genetic resources—in Lebanon and the Bekaa Valley, specifically. (These presentations included much of the information on climate changes in Lebanon summarized in chapter 2 and the impacts of these changes in the Bekaa Valley summarized in the section above). The second part of the workshop was the participatory process described below. The workshop included 65 participants, in addition to the local country organizing team from the Lebanese Agricultural Research Institute, World Bank staff, and several consultants. The participants were from several groups: farmers, LARI researchers and other staff, and representatives from Lebanese government ministries, national universities, and international organizations. Importantly, farmers were well-represented in the workshop, accounting for about 30 percent of the 65 participants. As their livelihoods depend on agriculture, farmers typically have the most detailed and comprehensive knowledge of "on the ground" climate changes and their effects on the biophysical environmental and agricultural production. For this reason, in prior studies similar to those reported here for Lebanon and Jordan, it has been recommended that farmers account for a significant proportion (at least 30 percent) of workshop participants (World Bank 2009).

Step 2, Workshop 1: Identification of Possible Response Options to Climate Changes

The second part of the workshop consisted of a participatory process in which workshop participants, based on the information at hand, identified a set of possible response options to address agricultural adaptations to climate change in the Bekaa Valley. A standard workshop-type format was followed, assisted by an experienced facilitator. Following the presentations on climate changes and their impacts, the participants were divided into several small groups (10–12 participants each) to "brainstorm" possible response options. The agricultural response options that were elicited in these sessions differed widely, reflecting the participants' diverse backgrounds, experiences, and knowledge of climate change impacts in the Bekaa Valley. After the small group discussions, the groups reconvened in a final plenary session to report their recommendations and discuss them with the entire group. The workshop facilitator assisted in reducing duplication of responses, achieving group consensus on the most important ones, and in organizing the suggested options into consistent categories for further analysis by the country team.

The climate response options identified by workshop participants are given below and vary widely. They range from technical measures pertaining to production agriculture and irrigation management, to a range of recommended investments in agricultural research, and possible legal and institutional changes designed to address resource constraints facing land, water and genetic resources. Measures related to water-related constraints, land-use and pest and disease management figured heavily in the concerns of farmers. Immediately following the workshop, the country team further reviewed these response options, and decided which ones were most consistent with the consensus views represented in the workshop as well as which options appeared most feasible for further review and analysis. These are discussed in more detail below.

Step 3. Workshop 2: Evaluation and Prioritization of Response Options for Climate Change Adaptation

The second workshop was held at the offices of the Lebanese Agricultural Research Institute in Zahle in the Bekaa Valley in January 2011. In addition to the workshop organizers and World Bank-related staff and consultants, a total of 70 individuals participated in the one-day workshop; many had attended the first workshop in October. Most of the institutions from government, research organizations, and the NGO community, as well as international institutions, which had participated in the first workshop were represented in the second. The main objective of this workshop was to involve the participants in the evaluation of selected possible response options that had emerged from the first workshop and to prioritize them for inclusion in a draft action plan to be developed subsequently. Several workshop presentations first summarized (from the first workshop) expected climate changes in Lebanon and likely impacts on agriculture in the Bekaa Valley region. After that, the workshop focused on three elements:

• Review of the selected priority response options that emerged from Step 2 (Workshop 1)

- Identification and weighting of the evaluation criteria to be used in the prioritization of the response options
- Scoring and prioritization of the response options, using the above evaluation criteria.

Each of these steps is here discussed in turn.
Climate Change Response Options Identified in Workshop 1: Lebanon

Agriculture:
- Maintain domestic assets during drought and for potential use in plant improvement programs in collaboration with international institutions to find potential varieties
- Use gene banks to support scientific research through collecting the largest possible quantity of genetic resources for sustainable use
- Identify and monitor appearances of emergent pests and diseases resulting from climate change, and employ new integrated pest management strategies
- Early warning systems for pests and diseases
- Promote adoption of drought-tolerant crop varieties under new climatic conditions
- Conserve and produce certified fruit tree rootstocks and varieties economically important and tolerant to drought.

Irrigation:
- Establish of mountain pools and medium-sized dams
- Rational use of irrigation water
- Re-use of wastewater
- Analyze irrigation water periodically
- Rehabilitate and establish basic distributed networks of irrigation water
- Legislate assuring legal use of irrigation water and the rationalization of the artesian well use
- Early warning system for irrigation.

Research and Extension:
- Agroclimatic characterization
- Research on good agricultural practices
- Research on modern irrigation methods
- Research on the use of alternative energy in agriculture
- Research on the quantification of chemical and organic fertilizer in different crops
- Adopt agricultural rotations in order to maintain the soil
- Activate extension and strengthen the relationship between the farmer and leaders
- Conduct economic feasibility of the plantations under the climatic conditions

- Publications and extension bulletins for farmers
- Renew hydrological maps identifying sites, and the ability to pump water from artesian wells in cooperation with the municipalities.

Legislation and Laws:
- Classify agricultural land
- Creat ex situ reserves to maintain genetic origins of drought-tolerant crops
- Pass emission laws for protected areas
- Introduce legislation and laws on the use of irrigation water
- International institutions to play an important role in support of the agricultural sector in light of climate change taking place
- Introduce mechanisms for dealing with disasters resulting from climate change and establishment of compensation to farmers.

Selection of Priority Response Options

Between the first and second workshops the country team narrowed the list of possible response options that had been identified by participants in Workshop 1 to a more selective list for further analysis and discussion by participants in Workshop 2. In this process, the country team took into account a variety of concerns, including the technical feasibility of each response option; the likelihood of each being able to address climate change adaptation in the Bekaa Valley; the resources potentially available, both from domestic and international sources; importantly, the extent to which the LARI could play a distinct role in executing the response option; and finally, the mix of technical, institutional, and other resources that could be brought to bear. The final set of response options—and the objectives of each—that was developed for discussion at the second workshop consisted of the following:

1. **Evaluation and Maintenance of Genetic Diversity** of wild species and local varieties adapted to climatic change in the Baalbeck-Hermel region, especially wheat, barley, *Prunus*, fig, and caper. Promote these species for cultivation in the region, improve plant genetics and produce new drought tolerant varieties.
2. **Production, Selection and Dissemination of Plant Materials Adapted to Climate Change:** Establish fruit tree species, including economically important varieties that are well adapted to climate change; promote the production and dissemination of certified plants with characteristics adaptable to climate change.
3. **Integrated Management of Pests, Diseases and Plant Disorders under Climate Change:** Identify emerging diseases and pests for main crops resulting from climate change in the Baalbeck–Hermel region; promote the development and application of integrated pest management to enhance productivity and sustainability; identify plant disorders under the current and expected climatic changes and develop appropriate solutions.

4. **Adoption of New Irrigation Technologies:** Decrease water use and increase water use efficiency and crop productivity by establishing better irrigation systems, particularly drip irrigation systems, improve scheduling of water applications (quantity and timing), and increase technical assistance in both.

5. **Constructing Small- and Medium-scale Water Harvesting Reservoirs:** Construct small ponds, tanks and water harvesting reservoirs for collection, storage, and utilization of water to increase agriculture production and reduce the depletion of water resources. Use of reservoirs to produce high quality certified potato seeds that minimize use of pesticides and promote use of IPM, in isolated mountain areas of northern Bekaa.

6. **Capacity Building for Climate Change Adaptation:** Increase technical knowledge and skills of LARI staff in targeted areas through improvement of quality and relevance of research; strengthen the agriculture faculty in technical and university institutions; facilitate communication between LARI and farmers through joint training sessions; formalize linkages with farmers to better communicate the results of research; undertake awareness-raising activities on climate change and impacts on farming practices

After the country team narrowed the list of response options to those listed above, short profiles of each of these response options were developed to present at Workshop 2. Each of the profiles summarized key elements of the response options—regarding objectives, proposed activities, costs, expected results, institutional arrangements, and a timeline—for discussion by workshop participants. A summary of the profiles is given in box 3.1.

Box 3.1

Response Option—Lebanon: Adoption of New Irrigation Technologies

Overview

Adapting agricultural practices to changes in climate in the Bekaa Valley will require various measures among which water-saving techniques are probably the most significant. Less water use in the agricultural sector will mean more water will be available for other human activities. In areas where water is a limiting factor for crop production, applying water saving techniques is a necessity. This is especially true with climate change threatening increased water scarcity in the region. Since agriculture is the dominant water user in this region and is characterized by low irrigation efficiency, improving the efficiency of irrigation systems is an important step to reforming water management systems. Installing higher-quality drip irrigation systems is one method that can help accomplish this.

Drip irrigation helps to increase irrigation efficiency and productivity by reducing the amount of water needed to successfully irrigate a given plot. This technology has demonstrated the highest levels of irrigation efficiency. Farmers of irrigated crops such as potatoes typically

(box continues on next page)

Box 2.1 Response Option—Lebanon: Adoption of New Irrigation Technologies *(continued)*

apply more than 1,000 millimeters of water per dunum,[1] but drip irrigation can reduce this by 30–40 percent. For this reason, drip irrigation is the preferred watering method in arid and semiarid regions. These systems are easy to design and install, and they help reduce disease problems associated with aboveground moisture. Through this technique, water is only applied to where it is most necessary—at the plant's roots. Drip irrigation systems can also be used to apply specific nutrients to the soils and crops.

Objective and Activities

The objective is to assist farmers in establishing drip irrigation systems. Activities would include (1) introducing demonstration drip irrigation systems in areas where there is little or no previous experience, (2) enhancing farmers' knowledge with regard to water application methods and techniques, (3) establishing demonstration plots under the supervision of local specialists (for example, the Lebanese Agriculture Research Institute (LARI)). This will provide practical experience in proper irrigation scheduling and water use.

Expected Results

1. **The installation of drip irrigation systems in an area of 1,000 dunum (about 100 hectares).** Small farmers will be primarily targeted. Lowering water consumption for irrigation will be the main effect of utilizing drip irrigation. Reducing the water application cost and increasing the water use efficiency will increase the farmer's economic returns. Farmers will pump water for fewer hours thus reducing their fuel-burning costs. Each farmer will save about US$200 to US$300 per dunum in irrigation techniques installation. In total, US$200,000 to US$300,000 will be saved for an area of 1,000 dunum.
2. Increasing crop yields through increased water use efficiency and the possibility of applying nutrients to address crop needs. Fertigation[2] techniques are one important portion of the new irrigation technologies that it is used for fertilizer application. Increasing water use efficiency means more yield for less amount of water application.
3. Increasing the effective contribution of research at the farm level. Demonstration fields on farmers' lands will help promote the adoption of the new irrigation techniques. LARI will also be involved in the preparation of booklets and brochures to help promote the new irrigation techniques.

Institutions and Partnerships

The introduction of new irrigation techniques at the farm level will require the involvement of research institutes such as LARI and the dissemination of information through the Ministry of Agriculture. Other civil society groups and municipalities would help in implementation.

Identification and Weighting of Evaluation Criteria

The second step in the priority-setting methodology entailed identifying the criteria by which the response options were to be evaluated and then weighting them for use in the subsequent step. The process of identifying evaluation criteria began following the first workshop, when the country team discussed a set of possible evaluation criteria to be used in evaluating the response options at the second workshop. The country team and external consultants engaged in an extended dialogue regarding these criteria, beginning with the use of criteria that had been used in a previous climate change adaptation project in Latin America (World Bank 2009). It was felt that these criteria matched the conditions in Lebanon fairly well, so these were used as the initial basis of discussion here. Input from participants at the first workshop and the country team's extensive knowledge of climate, agricultural conditions, and local communities in the Bekaa Valley were also used in identifying relevant evaluation criteria.

Based on this analysis, a draft list of *impact criteria* (evaluating the local impacts of climate change on agriculture in the Bekaa Valley) and *viability criteria* (assessing the viability of possible response options) was presented to the participants at the beginning of Workshop 2 for review, discussion and possible revisions. Following this discussion and in concurrence with the criteria proposed, the participants engaged in a criteria weighting exercise in which they were each asked to allocate 100 points among seven impact criteria and another 100 points across six viability criteria. Table 3.3 displays the criteria in each of the two groups and the average weights calculated among the workshop

Table 3.3 Impact and Viability Criteria, and Average Weights Assigned by Workshop Participants Used in Priority Setting: Lebanon[3]

Rank	Impact criteria	Average weight (n = 46)
1	Potential net economic benefits of the response option (for example, benefits minus costs)	19.7
2	Potential to promote adaptation to climate change and adjust to its effects on agricultural crops, fruit trees, etc.	19.2
3	Other environmental impacts	15.4
4	Complementarity between public and private sectors	14.2
5	Importance of the activity to the poor and local communities	13.0
6	Time required to achieve positive results	10.4
7	Indirect and spillover effects on other sectors	8.6
Total		**100**

Rank	Viability criteria	Average weight (n = 46)
1	Technical viability (and safety) of the response option	19.0
2	Importance of public sector intervention of response option	17.2
3	Degree of public support of the response option	16.4
4	Availability and quality of information needed to assess implications	16.3
5	Compatibility with the national strategy on climate change	15.6
6	Level of preparedness at local and regional level to implement response option	15.5
Total		**100**

Source: World Bank data.

participants (note that for this weighting exercise, usable and complete responses were available for 46 workshop participants).

Prioritizing the Response Options

The last step in the prioritization process was that of engaging workshop participants in a priority-setting exercise, given the preceding information on (1) the selected response options identified above, and (2) the evaluation criteria selected and weighted above. The priority-setting exercise was conducted at the conclusion of the second workshop. Participants were given a matrix in which the six response options were presented along with each of the seven impact criteria and six viability criteria. Participants were asked to assess each response option by assigning a value from 1 to 10, based on their individual evaluation of the extent to which each criterion was effectively addressed by each response option. The average scores that workshop participants assigned to each response option for each criterion were then weighted by the criteria weights calculated above). The impact criteria were proportionately assigned 50 percent of the total score and the viability criteria were proportionately assigned the remaining 50 percent, so that the scores could be consistently tabulated. The total average weighted scores for all response options across all workshop participants were then tabulated and the resulting scores normalized to a maximum value of 100.

The final results of this scoring exercise are presented in table 3.4. Four of the six response options have final scores clustered between 65.0 and 70.0, with only the "Evaluation and maintenance of genetic diversity" (for example, for wild species and local varieties) ranked by participants substantially lower than the rest. Two response options—the development of improved irrigation technologies (score of 70.3) and the construction of small-scale water harvesting reservoirs (score of 69.2) clearly outrank the rest. Perhaps this should not be surprising since the workshop participants consistently raised water-related constraints, related to both quantity and quality, as the primary constraints facing **existing** production systems. With climate change expected to exacerbate

Table 3.4 Results of Scoring Process for Climate Change Adaptation Response Options: Lebanon

Rank	Response option	Final score (max = 100)
1	Development of improved irrigation technologies	70.3
2	Construction of small-scale water harvesting reservoirs	69.2
3	Integrated management of pests and diseases	66.8
4	Production/dissemination of crops & plants adapted to climate change	65.8
5	Capacity-building for climate change adaptation in agriculture	65.1
6	Evaluation and maintenance of genetic diversity	56.4

Source: World Bank data.

future water scarcity problems (as this was related to participants by workshop presenters), it is only logical that water availability was the top-ranked option.

Lebanon Action Plan

After the second workshop, the country team revised the prioritized response options in the form of a draft action plan that would subsequently be presented and discussed by regional and national policy makers, research administrators, donor organizations, and other organizations. The key elements developed in the draft action plan for the Bekaa Valley are listed, in priority order, as follows, and explained in greater detail in appendix F

1. Adopt new irrigation technologies.
2. Launch project to construct small- and medium-scale water harvesting reservoirs.
3. Integrate production management of pests, diseases and plant disorders under climate change.
4. Produce and distribute crops and plants adapted to climate change.
5. Increase capacity for climate change adaptation.
6. Evaluate and maintain genetic diversity of wild species and local varieties adapted to climatic change.

Adopt New Irrigation Technologies

Objectives and Proposed Activities. The main goal of this project is to assist farmers by promoting watersaving demonstration drip irrigation systems so as to decrease agriculture water use and increase water use efficiency in the Bekaa Valley. This entails disseminating experiences of the Lebanese Agriculture Research Institute (LARI) in using drip irrigation for crop production. The proposed activities aim to reduce the quantity of applied irrigation water, increase water use efficiency and farm productivity, and help farmers establish well-designed irrigation systems. Farmers will also be trained in how to apply alternative watersaving application techniques with regard to quantity and timing. This project will introduce drip irrigation systems in areas where there is little or no previous presence of these irrigation techniques. Farmers' knowledge will be enhanced with regard to water application methods and techniques through demonstration plots under the supervision of LARI. Farmers could choose to enter a copayment system for drip irrigation system with cost-sharing involved. NGOs, cooperative societies and municipalities would assist in securing collaboration and communications with farmers.

Expected Results. (1) installation of 100 hectares of demonstration drip irrigation systems; (2) lowering water applications by 30–40 percent per crop, increasing water use efficiency, and decreasing fuel use (for pumping); (3) increasing crop yields through increased water use efficiency and the possibility of applying nutrients for crop needs (fertigation techniques);

(4) demonstration fields on farmers' lands will help promote the adoption of the new irrigation techniques; (5) LARI will be involved in the preparation of booklets, brochures and farmer training activities to help promote the new irrigation techniques, and in the collaboration with other local organizations.

Institutions and Partnerships. Lebanese Agricultural Research Institute (LARI), Lebanon's Ministry of Agriculture, NGOs, cooperative societies, and municipalities.

Launch a Project to Construct Small- and Medium-Scale Water Harvesting Reservoirs

Objectives and Proposed Activities. Securing non-conventional water resources is the main objective of this project. Construction of small ponds or tanks and medium-sized water harvesting reservoirs as a pilot project is proposed as an alternative means of collecting water in mountainous and plain-type areas. Water harvesting in Lebanon is a high priority, as it will help diminish the effects of climate change and conserve scarce water resources. In addition, small water reservoirs will provide gravity-fed water to downstream land, thus decreasing greenhouse gas emissions from pumping water wells. One of the main benefits of the water harvesting reservoirs in remote areas will be the production high quality (super elite and elite) potatoes. Affordable certified potato seeds will be produced that minimize the use of pesticides and that promote the practice of integrated pest management.

Expected Results. (1) increasing water availability, especially in the dry season, and introduction of supplementary irrigation to formerly rainfed areas; (2) rainwater harvesting reservoirs will improve cropping intensity and yields; (3) farmers will more likely rely on cheaper harvested water for irrigation, lessening the reliance on water wells; (4) expected increases in farmers' average annual net incomes by 50 percent from crops and livestock following reservoir construction; (5) construction of reservoirs in remote high-altitude areas will open the road to producing certified seeds for potato growers.

Institutions and Partnerships Lebanese. Agricultural Research Institute (LARI), Lebanon's Ministry of Agriculture, the Remote Sensing Center (National Council for Scientific Research), and NGOs.

Integrate the Production Management of Pests, Diseases, and Plant Disorders under Climate Change

Objectives and Proposed Activities. The goal of this project is to study the impact of climate change on plant diseases, pests and physiological disorders in the Baalbeck–Hermel region to help farmers better adapt to current and future climate changes. Observations, monitoring, and field inspections under climate change are important in order to assess the situation of the region and the phytosanitary status of the main crops. The focus would be on identifying emerging diseases and pests on important crops such as almond, apricot, cherry, apple, pear, quince and potato, etc. resulting from climate change; monitoring the development of these diseases and pests; and enhancing the productivity of cropping systems in the region through the development and application of sustainable integrated pest management (IPM). This will be done in conjunction with promoting the application of Best Management Practices (BMPs) among farmers in the demonstration fields, such as training and pruning of fruit trees, tillage, rational fertilization and fertigation, flower and fruit thinning, use of insect traps, etc. In addition, the identification of non-pathological plant disorders under current and expected climatic changes—including those resulting from drought and severe weather damage, salinity stress, and a variety of nutrient deficiencies—make the development of appropriate solutions critical for a comprehensive study.

Expected Results. (1) identification of emerging plant diseases, pests and physiology disorders in Baalbeck-Hermel region; (2) increased adoption of integrated pest management practices, providing technical assistance for farmers in the field of plant protection; (3) potential yield increases of at least 5–10 percent; (4) reduction in the indiscriminate spraying of agricultural pesticides of around 20 percent; (5) help identify solutions for more than 50 percent of plant health problems faced by farmers.

Institutions and Partnerships. LARI Department of Plant Protection; Department of Irrigation; local farmers and farmer organizations.

Produce and Distribute Crops and Plants Adapted to Climate Change.
Objectives and Proposed Activities. The production and delivery of certified plant species and varieties adapted to climate change can be one of the most economically productive strategies to adapt to climate change. There are two objectives in this proposed response option. (1) Introduce internationally important rootstocks and varieties known for their drought tolerance to assess their chilling requirements, flowering periods, resistance to physiological disorders, tolerance to pests and diseases, and other likely climate change impacts. (2) Select locally economically important rootstocks and varieties (such as the local almond and *Prunus Mahaleb* rootstocks) known for their adaptability to climate change, including drought tolerance and resistance to some pests and diseases.

Expected Results. (1) reduction of water use in irrigation; (2) improvement in fruit production and yields; (3) improvements in fruit quality and value; (4) expanded shelf life of fruit products; (5) opening new markets.

Institutions and Partnerships. Machatel Loubnan Nursery Association, private nurseries, farmers in Baalbeck-Hermel region, Lebanese Agricultural Research

Institute (Biotechnology Department, Plant Protection Department), Ministry of Agriculture's Plant Certification Department.

Increase Capacity for Climate Change Adaptation

Objectives and Proposed Activities. The objectives of this response option are to increase the technical knowledge and skills of the Lebanese Agricultural Research Institute (LARI) staff and to improve the quality and relevance of research conducted by LARI; to improve in-house knowledge and capacity to respond to climate risks through academic-level coursework; to strengthen collaborations with agriculture faculty in technical and university institutions; to facilitate communication between LARI and farmers through joint training sessions; and to improve the national institutional capacity for climate risk management working with various national and regional institutions. Communicating the results of research and investigations to smallholders requires improved formal linkages and improved training programs between LARI and farmers, undertaking awareness-raising activities such as workshops on climate change and impacts on farming practices, and improving skills and knowledge on climate change-related issues through training workshops. Other activities include: screening and revising the national development plan and rural development strategies; working with the Ministry of Agriculture on embedding climate adaptation plans in Ministry of Agriculture activities; establishing a national climate change authority, including a monitoring system to recommend adaptation and mitigation measures, and a database to improve how farmers can adapt to changes in the climate (floods, droughts, etc.); and developing early warning system(s) to provide daily weather predictions and seasonal forecasts aimed at reducing impacts and assisting climate change decision making.

Expected Results. (1) improved training of farmers and increased adoption of modern irrigation methods (like drip irrigation), enabling them to produce at commercial levels adapting to climate change effects; (2) research focused on Lebanon's climate change effects in agriculture, including the dissemination of LARI's experiences, and improved communications between scientists and farmers on climate change adaptation measures; (3) strengthened researchers', technicians' and staff awareness skills and expertise in climate change, which will be reflected at the farmer level; (4) improve farmers' ability to address climate change impacts and to improve the productivity and quality of their agricultural production through use of drought-tolerant and other adapted crop varieties, improved irrigation practices, etc.

Institutions and Partnerships. The Lebanese Agriculture Research Institute (Departments of Plant Biotechnology, Irrigation and Agrometeorology, Plant Protection, Plant Breeding, and the Central Laboratory); the Ministry of Agriculture (Extension Department); INRA (Institut National de la Recherche

Agronomique-France); CIHEAM–Bari (Centre International des Hautes Etudes Agronomiques Mediterranéennes); CIMA Foundation (International Centre For Environmental Monitoring, Italy); Saint Joseph University.

Evaluate and Maintain Genetic Diversity of Wild Species and Local Varieties Adapted to Climatic Change.

Objectives and Proposed Activities. Confronting climate change in agriculture must allow for the conservation and sustainable use of genetic resources. One mechanism is to maintain genetic diversity of the dominant wild species and local varieties adapted to climatic change. Key crops are wheat, barley, *Prunus*, fig, and caper. These species could be utilized for cultivation in the region or introduced through LARI's genetic improvement programs with farmer participation to produce new certified drought-tolerant varieties. Specific activities would include: (1) Survey, collect and define the local varieties and wild species of these crops; (2) Assess genetic diversity by using morphological and molecular markers (SSR technique); (3) Conserve the distinct accessions *in-situ* (on-farm) and *ex-situ* (gene bank); (4) Evaluate the drought tolerance of local varieties by using in-vitro techniques; (5) Introduce wild and local varieties in plant improvement programs on-farm and through field trials with farmers. Selected varieties would be adapted and produced as certified drought-tolerant material by LARI in order to distribute them to the farmers at low prices.

Expected Results. (1) improved, more productive, and more diversified crop varieties, resulting in higher production and increased farmers' incomes; (2) conservation of genetic diversity to address future climatic changes.

Institutions and Partnerships. Lebanese Agricultural Research Institute (Departments of Plant Biotechnology, Irrigation, Plant Protection, and Plant Breeding); ICARDA (International Center for Agricultural Research in Dry Areas), CNRS (National Council for Scientific Research), ACSAD (Arab Center for the Studies of Arid Zones and Dry Lands).Lebanese Agricultural Research Institute (Departments of Plant Biotechnology, Irrigation, Plant Protection, and Plant Breeding); ICARDA (International Center for Agricultural Research in Dry Areas), CNRS (National Council for Scientific Research), ACSAD (Arab Center for the Studies of Arid Zones and Dry Lands).

Jordan

Agriculture in Jordan and the Jordan River Valley

The agricultural sector in Jordan represented nearly 3 percent of GDP in 2010 (World Bank 2012), down from 8 percent in 1990 (Bani Hani 1996). The change is accounted for by the rise of the industrial and service sectors, which today represent 30 percent and 65 percent of GDP, respectively (CIA 2012). Over the last decade, Jordan had the largest decline in agricultural value-added

(the value added to agricultural commodities through processing and manufacturing) in the Middle East and North Africa region after only the West Bank and
Gaza. The agricultural sector represents the main livelihood for 20 percent of
the population in Jordan and represents 9.8 percent of its economically active
labor force, 70 percent of whom are women (FAO 2008).

Only 11 percent of Jordan's land is arable, and a much smaller proportion
hectares, 0.9 percent, is in permanent cropland (World Bank 2012). The highlands have 197,000 hectares of cultivated land area whereas the Jordan River
Valley has 33,000 hectares. Between 2003 and 2008, acreage has decreased in
the highlands and increased in the Jordan River Valley (tables 3.5 and 3.6).
Jordan's agricultural activity is concentrated in the Jordan Rift Valley, which
extends over 6,833 square kilometers at elevations ranging from +1,000 to −410
meters above sea level. The Jordan Rift Valley has two distinct agricultural zones
located in the north of the country: the mostly irrigated Jordan River Valley and
the mostly rain-fed highlands, while the drier eastern parts of the country are
home to the majority of the sheep and goat herding (Jordan Embassy 2012).
The Jordan River Valley is located in the northern part of the Rift Valley, and is
divided into three main parts: northern, central and southern. Approximately 40
percent of the country's population resides in the Jordan River Valley, which
produces above 50 percent of all agricultural output.

The Jordan River Valley has a sub-tropical arid and semi-arid climate with
mild winters and very hot summers, and receives more than 350 millimeters of
precipitation every year. The middle part of the Jordan River Valley around the
Dead Sea receives approximately 200 millimeters of rain a year, while the
southern part towards the Red Sea receives less than 50 millimeters per year.

Table 3.5 Irrigated and Non-Irrigated Crop Area in the Jordan River Valley (Hectares), 2003, 2008

Crop	Non-irrigated area (ha.)		Irrigated area (ha.)		Total (ha.)	
	2003	2008	2003	2008	2003	2008
Fruit Trees	22	137.9	9,084.6	10,100.6	9,106.6	10,238.6
Field Crops	911.8	306.4	2,016.2	3,038.6	2,928.6	2,345.1
Vegetables	12	0	13,906.7	20,714.1	13,918.7	20,714.1
Total	**945.8**	**444.3**	**25,007.5**	**33,853.3**	**25,953.9**	**33,297.8**

Source: Al-Naber 2010.

Table 3.6 Irrigated and Non-Irrigated Crop Area in the Highlands (Hectares) in 2003, 2008

	Non-irrigated area (ha.)		Irrigated area (ha.)		Total (ha.)	
Crop	2003	2008	2003	2008	2003	2008
Fruit trees	52,540	37,841	24,090	33,806	76,630	71,647
Field crops	127,890	98,602	4,150	5,686	132,040	104,287
Vegetables	2,290	1,664	16,410	19,492	18,700	21,156
Total	182,720	138,107	44,650	58,983	227,370	197,090

Source: Al-Naber 2010.

Temperatures in the Jordan River Valley can rise to 45°C in summer, and the mean annual temperature is 24°C. In winter, the temperature falls to a few degrees above zero, and frost is rare. Average temperatures in the Jordan River Valley are 6–9°C warmer than in the highlands (Al-Naber 2010). The rainy season extends from October to April, with the peak of precipitation taking place during January and February. Non-saline soils are typical in the northern parts of the Valley whereas the southern Valley is characterized by gypseous soils. Soils in the Middle Jordan River Valley are salty and sandy with excess calcium and a lack of iron (Venot 2003).

Jordan's agricultural production in concentrated in the Jordan River Valley. A variety of crops are found in this region. The northern part of the Jordan River Valley is predominantly citrus crops (57 percent by area), the middle part is predominantly vegetable production (56 percent), and the southern part is largely vegetable and banana production (70 percent) (Al-Naber 2010). The Highlands of Jordan are also productive areas. Zarqa, the eastern desert, and the Upper Yarmouk Basin each allocate between 66 and 70 percent of their acreage to vegetables, melons, and olives, and the rest to other fruit trees (apples and peaches) and seasonal crops (barley, wheat and alfalfa). Farms in the Highlands are typically large plots, averaging between 15 and 30 hectares per farmer.

The average farm size in the Jordan River Valley is 3.5 hectares, slightly above the national average of 3.3 hectares (IWMI 2012). Traditional inheritance laws have contributed to a decrease in the size of landholdings over time, although the government prohibits subdivisions of landholdings under three hectares (EAT/USAID 2012). Competition over land is intense. Molle, Venot, and Hassan (2008) observe that farms in the Jordan River Valley are highly hetero-geneous—they classify farms in the Jordan River Valley into five classes based on size, initial investment and net profits. These include: (1) family farmers, who either own or rent land (3–6 hectares) and grow vegetables in open fields; (2) entrepreneurial farmers, farming 6–10 hectares, using capital-, knowledge-, and labor-intensive techniques (for example, greenhouses) and who earn a high return on investment; (3) citrus orchards, located in the northern Jordan River Valley and operated by owners (with 3–6 hectares) or by managers hired by absentee investors (farming 1–20 hectares); (4) banana farms in the northern valley, which are 1–5 hectares in size and highly profitable; and (5) mixed farms of 1–3 hectares, with more extensive vegetable cultivation combined with small orchards (the poorest category of farmers). The principal sources of differences among these different farming systems lie in the reliance on capital use, the intensity of production, the type of land tenure, the irrigation technology used, and whether management is by owners or tenants. Incomes vary markedly, primarily depending on the type of irrigation system used. In an earlier study, Venot (2003) reports that in general, lower initial investment and higher net profits characterize the Jordan River Valley when compared with the Highlands, where farmers are poorer, and crops are rainfed and are grown extensively on small pieces of land (Venot 2003).

Most acreage in the Northern Jordan River Valley is dedicated to citrus (57.5 percent of area), vegetables (17.5 percent), barley and wheat (14 percent). Similarly, in the Middle Jordan River Valley, vegetables (56 percent), citrus (16 percent), barley and wheat (12 percent) take up most of the acreage. Land in the Southern Jordan River Valley is also dedicated mostly to vegetables (37 percent) and bananas (33.5 percent). In the highlands, owing to the rain-fed nature of agriculture, the majority of the acreage is dedicated to olive trees (57 percent) and grapes (9 percent) with much the rest being allocated to apple trees (5 percent) and vegetables (5 percent). Farms in the Highlands are characterized by their relatively large size (15–30 hectares on average) (Venot 2003).

Livestock production is an important part of the agricultural economy in Jordan, and is estimated to contribute about one-half of total agricultural GDP. Livestock production in 2008 amounted to a total of 28 mt of meat overall, including 5.8 mt of beef, 0.35 mt of camel meat, 15.7 metric tons (mt) of lamb, and 6.1 mt of goat meat. Additionally, 134.6 mt of poultry, 975.4 million eggs, and 330.3 mt of milk were produced. It is important to note that non-farm income is also an important contributor to rural income in Jordan, contributing as much as one-half (50 percent) (Adams 2001).

Jordan is heavily reliant on agricultural imports and cereal food aid. Although the country is self-sufficient in producing key commodities like fresh meat, eggs, tomatoes and olives, it only produces 0.7–1.5 percent of its total demand for wheat and barley (table 3.7) (Al-Naber 2010). Additionally, it is self-sufficient in fruits and vegetables and produces more than half of its domestic requirements of dairy and meat products. Vegetable production in Jordan is much higher than internal consumption, resulting in a large surplus available for agricultural export. Processed foods and beverages are also important exports (table 3.7).

The Jordan River Valley in intensively farmed, relying on irrigation for 98 percent of its agricultural production compared to a national average of 43 percent (DOS 2012). Irrigation is mainly supplied through a publically managed canal system, but private agricultural wells also exist. The irrigation system has been developed since the 1950s and 1960s, using surface water from the Yarmouk River and other secondary rivers (the "Side *Wadis*") (Al-Naber 2010). Around 65 percent of all water use and more than 50 percent of groundwater use in Jordan is withdrawn solely for irrigation purposes (Shatanawi et al. 2005). The total irrigated area in Jordan is estimated at 76,000 hectares; in addition to the 33,000 hectares located in the Jordan River Valley, the remaining area is in the highlands and the desert areas. The most common on-farm irrigation system is micro-irrigation, which is used in over 60 percent of the Jordan River Valley. Only a few farms use sprinkler irrigation, while the rest (about 35 percent) still use the more conventional, less water efficient surface irrigation method for citrus and banana cultivation. Overall, on-farm water efficiency is only 30–50 percent (EAT/USAID 2012). Molle, Venot, and Hassan (2008) suggest that

Table 3.7 Major Agricultural Production Sectors, Jordan, 2010

Top agricultural commodity production in Jordan

Rank	Commodity	Production (US$1,000)	Production (MT)
1	Tomatoes	272,465	737,261
2	Indigenous chicken meat	222,168	155,972
3	Olives	137,458	171,672
4	Cow milk, whole, fresh	79,014	253,200
5	Hen eggs, in shell	38,898	46,900

Top agricultural commodity exports from Jordan, by value, 2009

Rank	Commodity	Quantity (tons)	Value (US$1,000)	Unit value (US$/ton)
1	Tomatoes	431,713	169,004	391
2	Cucumbers and gherkins	98,688	69,811	707
3	Food Prep Nes	21,622	57,386	2,654
4	Beverage Non-Alc	104,256	43,231	415
5	Eggplants (aubergines)	83,549	40,335	483

Top 5 agricultural commodity imports to Jordan, by value, 2009

Rank	Commodity	Quantity (tons)	Value (US$1,000)	Unit value (US$/ton)
1	Rice Milled	166,161	172,521	1,038
2	Maize	522,414	154,246	295
3	Food Prep Nes	35,040	134,064	3,826
4	Sugar Refined	214,093	114,787	536
5	Wheat	519,313	109,317	211

Source: FAO 2010.

irrigation efficiency in the Jordan River Valley is suboptimal for many reasons, including "unstable pressure in collective pressurized [irrigation] networks," the prevalence of a traditional pricing system of block-rate tariffs based on crop-based quotas, problems of filtration and clogging, non-uniformity of water application, poorly designed irrigation block layouts and water use rotations, and the lack of system-level storage capacity.

In 2004, the total irrigated land in Jordan for agriculture was 99,029 hectares. Seventy-one percent of this land is located in the Jordan River Valley and 29 percent in the Highlands (FAO 2008). Major irrigated crops in Jordan include field crops (cereals), vegetables (tomatoes, cucumber, squash, eggplants, pepper, cabbage, cauliflower, and potatoes), and fruit trees (olives). Vegetables represent 69 percent of the total quantity of agricultural production and 42 percent of harvested irrigated land (Al-Naber 2010). In the Jordan River Valley fruits and vegetables are the main products; these are harvested from October to May, while in the higher elevations they are harvested from May to November (Jabarin 1997). Major products include tomato, eggplant, squash, cucumber, cabbage, cauliflower, and potato.

Water withdrawals in Jordan was estimated at 941 million cubic meters per year, of which almost 65 percent was for agricultural purposes, 31 percent for domestic use and 4 percent for industrial use (IFAD 2003). Current water use and projected future withdrawals are widely considered unsustainable when compared to aquifer recharge rates and river flows. This is the case in the Jordan River Basin, which is heavily dependent on its increasingly scarce water resources. The traditional flow of 620 million cubic meters irrigating 54,000 hectares in the Jordan River Valley is now down to 270 million cubic meters, enough to irrigate only 21,000–24,000 hectares (Al- Naber 2010). Much of the remaining land is equipped with irrigation systems but are not irrigated due to the lack of water. Demands on water resources have been estimated at nearly 40 percent above currently available supply (Al-Naber 2010). As a result, the area cultivated fluctuates from year to year, based on precipitation and irrigation supplies. In the Middle Jordan River Valley, the water used in agriculture is more saline (ranging between 800 and E1,200 parts per million) and more polluted than in the Northern Jordan River Valley (Al- Naber 2010; Venot 2003). With the overexploitation of the scarce water resources in the Jordan Rift Valley and the more frequent droughts in the last few years, it is expected that the Jordanian agricultural system will change (Saba, Al-Naber, and Mohawesh 2010). Haddad (2009) argues that food supply will be in shortage and poverty increased even if Jordan decreases its water withdrawal, unless this decrease is accompanied by investment to improve irrigation techniques and water management.

Poverty in Jordan, as in Lebanon, is primarily an urban phenomenon and is lower than in other MENA countries. However, a full quarter of the people living below the national poverty line are rural (IFAD 2003). Rural poverty is concentrated in the low rainfall areas of southern and eastern Jordan (IFAD 2002). It is not surprising then that rural people in Jordan are suffering disproportionately from chronic undernutrition. This is evidenced by rising stunting rates in children (under age 5) that reach 27.3 percent in rural areas compared to 15.8 percent in the cities (IFAD 2003). This undernutrition can impact human development and, as a result, can impact labor productivity. Despite Jordan being around the regional average in agricultural labor productivity, it lags behind neighboring countries such as Lebanon, Saudi Arabia, and the Syrian Arab Republic. Low agricultural labor productivity in Jordan is not recent a recent development and has been attributed to weather-caused fluctuations in output, low investments in the sector, relatively obsolete agricultural technology, and certain market barriers (Bani Hani 1996).

Impacts of Climate Change on Agricultural Systems in the Jordan River Valley

As reviewed above, climate changes in Jordan, and in the Jordan River Valley specifically, are expected to be significant by the middle of the twenty-first century. While, to some extent, climate impacts on agricultural production will be attenuated in irrigated regions, water deficits can be expected to have impacts on irrigated regions as well, and increased risk and uncertainty can be expected to characterize the overall production environment facing agriculture. These factors emphasize the importance of viewing climate impacts in an integrated, regional manner. In this section, we examine a number of the likely impacts of climate change on agriculture in the Jordan River Valley in coming decades. These impacts provide the backdrop for the climate adaptation response options identified and prioritized in the succeeding section of this chapter and that comprise the regional action plan for the Jordan River Valley elaborated below.

Crop production in Jordan is very climate sensitive, which is reflected in a high variability in crop production. Souab (2010) notes that the average ratio of harvested to cultivated areas in the period of 1996–2006 was only 68 percent for wheat and 44 percent for barley, implying significant climatic risk facing farmers. In 2007, the most important crop production included 20.25 tons (mt) of wheat, 9.19 mt of maize, 205.7 mt of potatoes, 13.7 mt of barley, 46.3 mt of bananas, 77.8 mt of apples, 221.59 mt of olives, 173.6 mt of citrus fruit, and 793.2 mt of tomatoes. However, in 2009, barley production soared to 17.7 mt and wheat production fell to 12.3 mt. Total production area figures indicate an increase in the production of fruits and vegetables in both regions (tables 3.5 and 3.6) and a decrease in the production of field crops in the Jordan River Valley (table 3.5).

Temperature Effects

Predicted temperature increases for Amman are between 1.3°C and 1.5°C for the 2020s, and between 3.0°C and 4.0°C by the 2080s, based on the regional downscaling model (see chapter 2). Warming in the Jordan River Valley is expected to be within a similar range, and there is relatively high agreement on these trends between models (Christensen et al. 2007; Evans 2009). In addition, these increases are expected to be greatest in the summer months, particularly at night, and lesser, although still positive, in the winter months (Christensen et al. 2007).

Tomatoes are the largest single crop, by value, produced in Jordan, and by far comprise Jordan's largest agricultural export, reaching almost US$170 million in value in 2009 (table 3.7). Tomato production is strongly affected by temperature. At 27°C, increasing the mean daily temperature by 2°C results in a steep reduction in fruit set (up to ~50 percent reduction), total fruit numbers (up to ~75 percent reduction), and in fruit biomass per plant (up to ~75 percent reduction) (Peet, Willits, and Gardner 1997). Even short-term exposure to high temperatures (>32°C) can depress photosynthesis for some time afterward (Zhang et al. 2011). High temperatures can also affect tomato quality, which often reduces the marketable yield due to factors such as poor pigment development, sun scalding, or fruit cracking (Garg and Cheema 2011). Other important vegetable crops in the Jordan River Valley include cucumbers, potatoes, and peppers. It has been shown that using rootstock from heat-tolerant varieties, and even other species, may help promote resistance to temperature stress (Schwarz et al. 2010).

Interactions between temperature and precipitation changes can result in complex crop dynamics. In Jordan, summer temperature increases are predicted to be greater than increases in winter temperatures, which may reduce the relative impact of winter temperature changes as compared to those in summer. Still, some warming is predicted to occur during the winter, when moisture and rainfall tend to be high and temperatures are low, so one might expect increases in yield. However, the increase in temperature would be expected to reduce the length of the growing season, as discussed above, and so may shorten the grain filling period for field crops, thus reducing yields. Despite this, a faster growing season could also reduce crop water requirements, which could be beneficial for water resources (Khresat 2010).

The complexity of these crop dynamics makes modeling an attractive approach to answering some of these questions. Al-Bakri et al. (2010) ran a crop-climate model for Jordan's Second National Communication to the UNFCCC to investigate how predicted changes in rainfall and temperature would affect wheat and barley yields. They found that warming by 1–2°C would decrease barley yields by 14–28 percent, with negative impacts observed regardless of positive or negative changes in precipitation. However, for wheat crops, temperature increases of 1–2°C could increase yields, but only so long as precipitation remains constant or increases. In scenarios where precipitation decreases, the effects of an increase in temperature cannot compensate for the wheat yield decreases due to decreased moisture. Barley is currently planted in more marginal areas than wheat, and is thus more susceptible to changes in water availability. In addition, wheat is harvested in June, while barley is harvested in May, making barley more susceptible to the shorter grain-filling time resulting in lower yields (Al-Bakri et al. 2010). Thus, while both crops are vulnerable to climatic variations, barley, already planted in marginal regions and used for livestock fodder, may be particularly at risk. Both of these crops are critical for subsistence farmers' livelihoods.

Precipitation Effects

Agriculture in the Jordan River Valley already consumes 75 percent of the country's available water (Breisinger et al. 2010), and expansion of agricultural irrigation (along with industrial and tourism increases of water use) continues to increase these demands (Wilby 2010). The total area of arable land is strongly limited by water availability, whether for irrigation or from precipitation; the International Food Policy Research Institute predicts that per capita water availability in Jordan may decline by over 50 percent by 2050 (Breisinger et al. 2010). Changes in precipitation in the Jordan River Valley associated with climate change are less certain than for temperature, and are being imposed on already high inter-annual variability in precipitation. Nonetheless, precipitation in this region is predicted to drop by 5–15 millimeters by 2050 and 10–50 millimeters by 2095 (Evans 2009), and Wilby (2010) predicts even greater decreases for Amman (up to 105 millimeters by the 2080s). Importantly, this could push the region out of the range of precipitation suitable for rain-fed agriculture (200 millimeters per year) (Evans, Smith, and Oglesby 2004). Samuels et al. (2010) predict that base flows in the Jordan River Valley, which provides much of the Jordan River Valley's irrigation water, will decrease proportionally with the reduction in rainfall, while surface flow will decrease at an even greater rate. At the same time, the variability of water flow is predicted to increase, resulting in higher risks from extremely high flows as well. Decreasing the water available for irrigation in the Jordan River Valley could have serious impacts on productivity in the region, as could decreases in rain-fed agriculture in the Valley and in the nearby highlands. Climatic changes in other regions of the country could increase desertification and increase pressure on the remaining arable lands, leading to increased inputs and withdrawals of water.

Changes in precipitation may also influence whether largely rain-fed agriculture, such as olive trees and other tree fruits, is viable. While in the Valley most tree crops are irrigated, in the highlands over half of these crops are not irrigated. Moriondo, Stefanini, and Bindi (2008) modeled the range of habitat suitable for olive trees, and found that cumulative annual rainfall greater than 240 millimeters formed the boundary for suitable habitat. As areas of olive cultivation in Jordan already lie near this boundary line, predicted decreases in precipitation could likely make some areas unsuitable for rain-fed olive cultivation. Shatanawi, Al–Bakri, and Suleiman (2007) found that lemon yields in the central Jordan River Valley could be sustained under a deficit irrigation approach with a 25 percent reduction in total irrigation water use.

These results highlight the importance of agricultural research, the development of drought-resistant crop varieties, and improved crop production practices as adaptation responses to drier climates, anticipated water deficits and changing agroclimatic conditions. Successful crop breeding programs involving farmer participation have been conducted in Jordan (Fufa et al. 2010). In the work they describe, in most cases, farmers were more efficient than breeders at selecting traits that would result in high grain yields for their localities. The authors suggest that such programs are key to "increasing and stabilizing productivity in marginal environments as each specific environment is occupied by the best genotype" (Fufa et al. 2010).

CO_2 *Effects*

As discussed in box 2.1, potential CO_2 fertilization effects in C3 crops may be offset by water deficiencies. Key crops in the Jordan River Valley are predominantly C3 crops. While Leavitt et al. (2003) found that increased atmospheric CO_2 concentrations increased water use efficiency in sour orange trees in a growth chamber experiment, they did not study this effect in combination with temperature or moisture stresses. In a study of the interactions between CO_2 and soil moisture status, Valerio et al. (2011) studied the effect of CO_2 fertilization in a tomato (C3 plant) and *Amaranthus retroflexus* (pigweed—C4 plant) system with and without water stress. Under conditions with sufficient water, they found that the tomato plants responded more positively to the increased CO_2 than the weeds. However, when water was limiting, the tomato plant suffered more than the weed, even under increased CO_2, indicating that the negative effects of drought stress outweighed the positive effects of CO_2 fertilization for the tomato relative to the weed (Valerio et al. 2011). This study highlights the complexity of agricultural systems' responses to climate change and the need for further study of important cropping systems.

Pest and Pathogen Management

Key pests in Jordan include the olive bark beetle, red spider mites, the tomato leaf miner, and the South American tomato moth (Abdel-Wali 2010). Future climatic changes are widely expected to increase crop pest and disease problems, including the introduction of new pests and crop diseases as temperatures rise. Given that temperature increases in Jordan are expected to be particularly high at night, combined with decreased precipitation, this could conceivably decrease the success of fungal pathogens. Similarly, the prevalence of olive knot disease, present in Jordan and caused by the bacterium *Pseudomonas savastanoi*, tends to increase with increasing rainfall, and thus might be predicted to decrease as well (Teviotdale and Krueger 2004). However, warming can also decrease plant resistance to pathogens (Harvell et al. 2002) making net impacts difficult to predict as a result of these complex system interactions. Integrated Pest Management (IPM) has been introduced to Jordan, but it has largely been limited to tomato and cucumber crops (Johanson 2005). A wider application of this management strategy could be key in adequately controlling changing pest and pathogen pressures under future climate scenarios.

Effects on Livestock

While most livestock in Jordan are raised in the dry eastern regions, sheep, goats, poultry, and some cattle production still make an important contribution to agriculture in the Jordan River Valley and the highlands (Al-Naber 2010). Pastureland is widely overgrazed, so any reductions in viable pastureland (much of which is already marginal) associated with climate change could cause additional damage and further increase the intensity of overgrazing. In addition, feed resources are already limited, so any declines in crop production could have severe effects on the success of livestock operations. Interestingly, barley crops that fail during dry years are used as fodder, which might have somewhat of a mitigating effect on animal feed stresses during dry years (Al-Jaloudy 2006). Still, in bad drought years, in general, animal slaughter rates have gone up in Jordan (Al-Jaloudy 2006).

Animal diseases are currently a large cost to Jordanian farmers (Al-Jaloudy 2006). Increased stress on animals or changes to the vectors or agents of disease due to climate change, as with crops, could exacerbate animal disease and the associated losses (Harvell et al. 2002). The National Centre for Agricultural Research and Technology Transfer (NCARTT) has an integrated livestock program, which aims to support animal production in Jordan through initiatives including improving local breeds through selection and breeding, farm management, and animal feed quality, in addition to working to sustain and protect rangelands (Al-Jaloudy 2006).

Project Description and Results

As in the Bekaa Valley case discussed above, the project in Jordan revolved around the completion of four steps, encompassing two workshops and the development of a draft action plan for climate change in agriculture in the Jordan River Valley. Each step and the accomplishments in each are discussed briefly below.

Step 1, Workshop 1: Review of Climate Change and Impacts on Agriculture in the Jordan River Valley

The project workshops in Jordan followed the same structure outlined above in the methodology and which was followed in Lebanon. The first workshop was held in Amman in October 2010, and consisted of two parts. The first was a series of presentations reviewing the scientific literature and research evidence on (1) agriculture and natural resources in the Jordan River Valley, (2) climate changes in the Middle East, in general and in Jordan specifically, and (3) the impacts of climate change on agriculture, on resources—especially water resources—in Jordan and in the Jordan River Valley (summarized in the previous section). The second part of the workshop was the identification by workshop participants of a set of possible response options to address climate change adaptation in the Jordan River Valley. In the first workshop there were 66 participants, in addition to the country team led by NCARE, World Bank staff, and consultants. The participants were from several groups: farmers (17 participants); Jordanian government Ministries (7); universities (14); the National Center for Agricultural Research and Extension (13, plus local organizers); and national and international organizations, NGO's, and news media (15). As in Lebanon, the representation by farmers, who constituted more than 25 percent of total participants, was judged to be particularly important, given this project's objective of eliciting their input on climate adaptation strategies as those with first-hand experience with recent climate changes and their effects in agriculture. In addition to farmers, agriculture, water and climate researchers were well represented, coming from both the university community and NCARE.

Step 2, Workshop 1: Identification of Possible Response Options to Climate Changes

In the second part of the workshop, participants identified a set of possible response options to address needed agricultural adaptations to climate change in the Jordan River Valley. As in Lebanon, the format was a standard workshop format in which, after plenary presentations on agriculture, climate change and its agricultural impacts, participants broke out into several small groups to brainstorm possible response options. A wide diversity of responses was elicited, based on the participants' diverse backgrounds, training, familiarity with the region, and knowledge of climate change impacts in the Jordan River Valley. Following this, the groups reconvened in a final plenary session to report back their results to the group as a whole. The workshop facilitator assisted the process of eliminating duplication, grouping like responses and organizing the responses. Immediately following the workshop the country team reviewed the response options that arose during the workshop, and further narrowed down the response options to those that appeared to represent the main consensus views that arose in the workshop and that were deemed most feasible. The range of possible adaption response options that was identified, grouped into 15 areas, is listed below. Selected response options that were the subject of further analysis and consideration are discussed in more detail below.

Step 3, Workshop 2: Evaluation and Prioritization of Response Options for Climate Change Adaptation

The second workshop was held in Amman in January 2011. In addition to the workshop organizers and World Bank-related staff and consultants, 70 participants attended the workshop, many of whom had attended the first, and some of whom were new. Most of the government, research, NGO and international institutions, which had been represented in the first workshop, were again represented. The main objective of the second workshop was to engage the participants in the evaluation of the proposed response options that had emerged from the first workshop and to prioritize them for inclusion in a later action plan. After an initial summary review of expected climate changes in Jordan and likely impacts in agriculture in the Jordan River Valley region, the workshop focused on three aspects:

- Review of the priority response options that emerged from Workshop 1
- Identification and weighting of the evaluation criteria to be used in the prioritization of the response options and
- Scoring and prioritization of the response options, using the evaluation criteria in #1.

Each of these steps is here discussed in turn.

Climate Change Response Options Identified in Workshop 1: Jordan

1. Water resources management: to address water availability and quality
   - Water harvesting and storage, for irrigation
   - Improve water use efficiency
   - Improve management of water discharges. Use and develop improved water management technologies (for example, drip irrigation, magnetized water, etc.)
   - Improve technologies for water desalination (solar, nano, etc.)

- Improve irrigation practices
- Create new ethic for recognizing and coping with water scarcity
- Develop policies and programs to improve farm-level water and resource management
- Research on hydrologic cycle, water management, etc.

2. Farm production and management: improved efficiency and sustainability
   - Develop and use new crop varieties, including hybrids, especially those which maximize yields subject to drought and high temperatures, which economize on water and fertilizer use, and which address changing environmental conditions and associated risks.
   - Improve practices such as optimizing growing season, crop rotations to minimize pest and disease problems, crop and varietal diversification in response to changing environmental conditions and increased climatic risks.
   - Use integrated pest management.
   - Use fertigation, especially to reduce/optimize water and nitrogen use.
   - Intensify production to increase production efficiency.

3. Land-use:
   - Modify land-use law to limit loss of arable land to urbanization.
   - Change location of crop and livestock production to address changing environmental conditions and economic risks.
   - Use alternative fallow and tillage practices to address moisture and nutrient deficiencies induced by climatic changes.
   - Change land topography to address moisture deficiencies and reduce risk of land degradation.

4. Research & development: especially improved plant varieties, improved seeds, animal management improvements, and new farming systems better adapted to climate changes

5. Private and private linkages: strengthen links (with both national and international public and private entities); support/assist national efforts to deal with climate changes.

6. Capacity building: Develop skills, expertise and awareness among relevant institutions (public, private, NGO, research organizations, local communities) of the need to reduce greenhouse gases (GHGs) and of measures needed to reduce vulnerability to climate change impacts. Could include workshops, training for improved management, field visits, tours, etc.

7. Climate Change National Authority: Establish a national authority with responsibility to develop, evaluate and ensure implementation of adaptation strategies; such authority would have representation from relevant Ministries (Agriculture, Health, Energy, etc.)

8. Weather and climate information system: develop and improve existing early warning system to provide daily weather prediction and seasonal forecasts.

9. Climate Change Monitoring and Evaluation (M&E) system: Establish a national climate change M&E system to identify key climate risk areas recommend appropriate adaptation and mitigation measures, and evaluate progress. Would help in measuring the success of a climate change strategy and in ensuring that governmental service delivery incorporates climate change factors. Would require staff, equipment, facilities, and development of appropriate indicators.

10. Databases: Develop a database for use by farmers and local communities on how to best deal with climate events such as droughts and flooding, and how livelihoods, location, adaptation, and institutions affect outcomes.

11. Livestock production and marketing:
   • Encourage water harvesting for livestock use.
   • Establish and promote small dairy and meat businesses.
   • Identify alternative feedstuff resources, intensify livestock production (for example, fewer but higher producing animals), and improve livestock feeding efficiency, all to help reduce livestock methane emissions.
   • Diversify livestock types and breeds to address environmental constraints and risks.
   • Improve livestock management: pasture & feeding; plant drought-tolerant grasses; construct dikes and dispersed watering stations; etc.

12. Agricultural pests and diseases: Assess and evaluate risks of increased frequency and severity of pest and disease problems due to climate change:
   • Assess state of science and research needs regarding projected changes in disease vectors.
   • Assess threats from dormant and controlled diseases, pests and invasive species.
   • Develop adaptation and intervention strategies.
   • Reform policies, procedures, and staffing in response to new threats.

13. Climate change policies and funding: Develop adaptation and mitigation policies and measures into development plans, leading to improved adaptation and reduced GHGs, including:
   • **Natural Disaster Fund:** to help affected communities and farmers meet emergency needs for survival and recovery, and to transition back to long-term development objectives
   • **Crop insurance**: provide crop insurance options that farmers could purchase to reduce risk of weather and climate-related income loss
   • **Incentive programs**: to reduce emissions and preserve and enhance carbon sinks. Help provide impetus to reduce CO_2 emissions by investment in renewable energy and more efficient technology.

Selection of Priority Response Options

Between the first workshop (held in October, 2010) and the second, the country team narrowed the list of possible response options that had been identified by participants in Workshop 1 to a more selective list for further analysis and discussion by participants at Workshop 2. In this process, the country team took into account a variety of concerns, including the feasibility of each response option; the likelihood of each response option being able to address climate change adaptation in the Jordan River Valley; the resources potentially available, from both domestic and international sources, to support these interventions; the extent to which the National Centre for Agricultural Research and Extension (NCARE) could play a role in effectively addressing the response option; and finally, the mix of technical, institutional, and other resources and partnerships that could also be employed in each ease. The final set of response options—and the objectives of each—that was developed for discussion at the second workshop was as follows:

1. Building Capacity for Climate Change Adaptation in Sectors Related to Agriculture:
   • Develop capacity that enables farmers, researchers and decision makers to plan and implement climate change adaptation measures in the country.
   • Develop capacity to conduct research, promote documentation and information on climate change.

2. Crop Production and Productivity: Improve the production and productivity in the crop sector through planning and implementing adaptation and mitigation options to address climate change and variability.
3. Livestock Production and Productivity: Enhance the production and productivity of animal-based agriculture in the Jordan River Valley through appropriate research and other climate adaptation measures
4. Land-use Change and Diversification: Reduce risks associated with climate change and variability by implementing new land-use law and encouraging land-use changes, improved farming practices and diversification.
5. Farm Water Management: Improve water use efficiency in farm water management through improved water harvesting, irrigation techniques, groundwater management, water quality improvements, creation of water user associations, and other interventions.
6. Agricultural Pests and Diseases: Reduce the risk of new increases in the frequency and severity of outbreaks of disease, pests and weeks due to climate change, through research, pest and disease monitoring, increased use of integrated pest management (IPM), etc.
7. Feasibility Study for Drought and Crop Insurance Program in Jordan: conduct a feasibility study aimed at reviewing the existing national drought strategy, with the potential to introduce a comprehensive long-term drought preparedness plan and effective drought insurance schemes to Jordanian government, farmers, NGOs and rural micro-credit organizations.
8. Early Warning System to Provide Weather, Climate, and Crop Risk Information to Farmers: Reduce the risk and vulnerability to farmers associated with agricultural drought, other extreme weather conditions, and pest and disease outbreaks by setting up an improved early warning system in Jordan.

Box 3.2

Response Option—Jordan: Reinforce Early Warning System for Drought

Overview:
Following the establishment of the Jordan's Drought Monitoring Unit (2008), a drought early warning system (DEWS) was established in the National Center for Agricultural Research and Extension. This system monitors drought by identifying drought severity and its geographic distribution through the remotely sensed NDVI (normalized difference vegetation index) indicator, which uses MODIS[4] images and seasonal rainfall records. NDVI maps are produced every 16 days (depending on the availability of MODIS data).

Objectives:
The frequency and severity of droughts are expected to increase in the future, as a result of climate change. Accordingly, improved alternatives for drought monitoring and effectiveness need further attention.

Proposed Activities:
- Analyze different alternatives for remotely sensed indices of drought (vegetation condition index (VCI), temperature condition index (TCI), vegetation health index (VHI), etc.) on a real-time basis, and evaluate the attributes of each as applied to the Jordanian situation.
- Conduct field studies of drought effects and their correlations with different remotely sensed indices.
- Perform a socio-economic study for selected affected areas to identify the potential economic damages of drought (for example, yields), and how they correlate with different drought indices and indicators (NDVI maps, field vegetation status, rainfall patterns).

Expected Results:
The development of more effective methods and indicators to minimize the impacts of drought on people and agriculture. Long-term benefits include: increasing farm household incomes, improving agricultural sustainability, and enhancing rural community resilience as a result of an improved ability to respond to drought.

Institutions and Partnerships:
Jordan Meteorological Department; Ministry of Water and Irrigation Water, Soil and Environment; NCARE's Drought Monitoring Unit (DMU) and other relevant departments, including: Field Crops, Horticulture, Integrated Livestock and Rangeland, and Socio Economic Studies. International collaborators of the DMU including the Arab Center for the Study of Arid Zones and Dry Lands (ACSAD) and the World Food Programme.

Short profiles of each of the response options were developed by the country team prior to Workshop 2 and were presented at the workshop. Each of the profiles presented in summary fashion key information about aspects of the response option for review and discussion by workshop participants. The profiles contained information regarding the objectives of each response option, proposed activities, costs, expected results, institutional arrangements, and a timeline. A summary of one of the profiles is given in box 3.2.

Identification and Weighting of Evaluation Criteria

This step in the prioritization methodology involves first identifying and weighting the criteria by which the response options are subsequently to be evaluated. This process actually began at the first workshop, immediately after which the country team identified a set of possible evaluation criteria to be used in the second workshop. Between the first and second workshops, the country team and external consultants engaged in a dialogue regarding these criteria, attempting to narrow them down to a practical and manageable number for use in the second workshop. This dialogue included input from discussions with participants in the first workshop, use of the criteria that had been successfully used in similar prioritization exercises for climate change adaptation in a prior World Bank project in Latin America (World Bank 2009), and the country team's extensive knowledge of the Jordan River Valley region, its agricultural systems and its people.

Based on this work, at the beginning of Workshop 2, a draft list of **impact criteria** (evaluating the local impacts of climate change on agriculture in the Jordan River Valley) and **viability criteria** (assessing the viability of possible response options) was presented to the participants for review and discussion, and to offer participants the opportunity to make revisions in these criteria. Following this, workshop participants engaged in a criteria weighting exercise in which they were each asked to allocate 100 points among nine final impact criteria and another 100 points across six final viability criteria. Table 3.8 displays the final criteria in each group and the average weights calculated across the workshop participants (note that for the criteria weighting exercise, complete and usable responses were available for only 41 of the workshop participants).

Among the impact criteria, the net economic benefits (for example, benefits minus costs) of each response option were clearly identified as the top-ranked criterion (weight = 18.0). The capacity of each response option to "moderate or reduce the impacts of climate change" on agriculture, especially the expected damages resulting from drought, floods, and other extreme events, was ranked second (weight = 15.6). Impact criteria 3–8 (table 3.8) were ranked at roughly similar levels by participants. The likely spillover effects on other regions— namely, the potential of the response option to address climate impacts in other regions of Jordan (or beyond) was ranked the lowest. The viability criteria ranked the highest by workshop participants all pertained, in one way or another, to the availability of resources to assure the viability of the response

Table 3.8 Impact and Viability Criteria, and Average Weights Assigned by Workshop Participants Used in Priority Setting: Jordan

Rank	Impact criteria	Average weight (n = 41)
1	Potential net economic benefits of the response option (for example, benefits minus costs).	18.0
2	Capacity to moderate or reduce impacts of climate change, including damage from extreme events (lower rainfall, drought, floods, etc.) to crop, livestock and horticultural sector.	15.6
3	Time required to achieve positive impact of the selected response option.	10.6
4	Potential to improve welfare of the poor and other vulnerable groups.	10.2
5	Flexibility of the response/adaptation option: Is the strategy reasonable for the entire range of possible changes in temperatures, precipitation, and altitudes?	9.8
6	Other environmental effects: soils, biodiversity, etc.	9.7
7	Private v. Public Sector: Does the strategy minimize governmental interference with decisions best made by the private sector?	9.7
8	Overall potential of the response option to address climate change impacts.	9.4
9	Spillover effects in other regions and sectors.	7.6
Total		**100**

Rank	Viability criteria	Average weight (n = 41)
1	Availability of money and credit resources to invest in the response option.	22.5
2	Adequacy of productive resources (water, land, etc.) to ensure viability of response option.	18.7
3	The availability of dependable information to implement the response option—databases on climate change, scientific information, monitoring and evaluation systems, etc.	18.1
4	The availability of educated and skilled labor/extension personnel, institutional capacity, appropriate facilities and equipment, and adequate infrastructure to implement the response option.	16.8
5	Acceptability of the response option to the public.	12.9
6	Institutional and legal feasibility of implementation.	11.0
Total		**100**

Source: World Bank data.

option. The importance of these resources was ranked as follows (in order): money and credit; productive resources (water, land, etc.); dependable scientific and other information to enable implementation; the availability of personnel, institutions, facilities and infrastructure to assure viability; acceptability of the response option to the public, and finally, institutional and legal feasibility of implementation.

Prioritizing the Response Options
The final step in the prioritization process was that of engaging the workshop participants in a priority-setting exercise given the information above on (1) the response options presented and discussed in part (a) above, and

(2) the evaluation criteria identified and weighted in part (b). At the conclusion of the workshop, all participants were given a matrix in which the nine response options were presented along with each of the nine impact criteria and six viability criteria. Participants were asked to evaluate each response option by assigning a value from 1 to 10 based on their individual assessment of the extent to which each criterion was effectively addressed by each response option. The average scores assigned by participants to each response option for each criterion were then weighted by the criteria weights previously calculated in part (b). In this process, the impact criteria were proportionately assigned 50 percent of the total score, and the viability criteria were proportionately assigned the remaining 50 percent. The scores were then calculated for all response options across all workshop participants and the resulting scores normalized to a maximum value of 100.

The results of this scoring exercise are presented in table 3.9. The results show a continuous range of scores across the nine alternatives. Participants clearly identified three response options having to do with increasing agricultural production—of crops and livestock—and improving water use efficiency through a variety of technological, management and institutional changes as the major priorities. This is not unexpected, given the severe constraints on water availability in the Jordan River Valley and, at least for farmers, their dominant concerns with raising production levels in order to increase (or even maintain) their income levels. The response options prioritized in this exercise subsequently figured prominently in the action plan for climate change adaptation in the Jordan River Valley developed following the workshop.

Jordan Action Plan

Following the second workshop, the prioritized response options were revised by the country team in the form of a draft action plan, with the intention of later presenting this to regional and national policy makers, research administrators, and donor organizations for their review and consideration. The key elements developed in the draft action plan for the Jordan River Valley are

Table 3.9 Results of Scoring Process for Climate Change Adaptation Response Options: Jordan River Valley

Rank	Response option	Final score (max = 100)
1	Increase farm production and productivity	72.0
2	Increase water efficiency	71.3
3	Increase livestock production and productivity	69.9
4	Capacity-building for climate change adaptation	65.8
5	Develop national climate change strategy	65.4
6	Reduce risks of agricultural pests and diseases	63.9
7	Establish climate early warning system	59.6
8	Implement land-use law and foster land-use changes	57.6
9	Feasibility study for drought and crop insurance	54.4

Source: World Bank data.

listed, prioritized and briefly described below. These are also summarized in a table in appendix G.

1. Improve farm production systems and productivity
2. Improve on-farm water use efficiency and integrated water resources management
3. Improve livestock and rangeland systems
4. Build national capacity for climate change adaptation
5. Reduce risks of agricultural pests and diseases
6. Reinforce early warning system for drought
7. Reform land-use laws and implement sustainable land-use
8. Activation of agricultural risk management fund.

Improve Farm Production Systems and Productivity

Objectives and Proposed Activities. The part of the action plan that was ranked highest by project stakeholders was to invest is a set of integrated crop production-related activities. These focus on increasing agricultural productivity in the horticultural and cereal sectors in the Jordan River Valley by enhancing the capacity of farmers to respond to anticipated climate changes. These activities are both generalized and commodity-specific. Proposed activities include: (1) evaluating and introducing new crop varieties (vegetables, tree crops, cereals) that respond to anticipated climate changes; (2) identifying and introducing alternative cropping patterns and cultural practices capable of better withstanding extremes of weather, drought, flooding and variable moisture availability; (3) increasing and sustaining wheat production in dry areas through (a) collecting seeds and increasing seed production of wheat landraces that are effectively adapted to climate change, and (b) establishing, promoting and disseminating on-farm field conservation agriculture practices; (4) collecting, conserving and utilizing wild barley genetic resources to address climatic constraints in barley production; and (5) conserving genetic diversity of major medicinal plants *in situ* and *ex situ* and their sustainable utilization to help diversify farmers' production alternatives.

Expected Results. (1) introduction of new crops and varieties for fruits and vegetables better adapted to climate change; (2) evaluation of existing cropping systems (wheat, barley) to identify key constraints and potentials; (3) identification of wheat landraces and genotypes which thrive in no-till systems; (4) identification and utilization of wild barley accessions which improve barley production under increased heat and low moisture conditions; and (5) conservation, characterization and utilization of threatened plant species, including major medicinal plants.

Institutions and Partnerships. National Center for Agricultural Research and Extension (NCARE)—various departments; Drought Monitoring Unit, Ministry of Agriculture; Jordan Meteorological Department; University of Jordan;

International Center for Agricultural Research in Dry Areas (ICARDA); Ministry of Education; local agricultural schools; and NGOs.

Improve On-Farm Water Use Efficiency and Integrated Water Resources Management

Objectives and Proposed Activities. Improvements in water use efficiency and water management in agriculture were consistently ranked as highly important by project stakeholders. To these ends, a number of innovations, research and feasibility projects are proposed, including: (1) the promotion of rainwater harvesting technologies and practices; (2) the design of improved wastewater treatment systems for water reuse in the production of selected crops (principally fodder and cut flowers); (3) evaluating the economic feasibility and environmental impacts of using treated greywater for irrigation; and (4) the evaluation of the potential for deficit irrigation in improving water productivity of vegetable crops. Other activities could include improved use of brackish water and the introduction of strict monitoring systems for groundwater exploitation in key locations.

Expected Results. (1) distribution of rainwater harvesting systems; (2) economically and technically feasible operation of wastewater treatment systems; (3) sustainable and economically efficient use of scarce water resources, including greywater, in agricultural production; and (4) determining the feasibility of deficit irrigation water management practices. As a result of all these steps, improved water use efficiency and improved livelihoods of farmers both in the Jordan River Valley and more marginal landscapes (Badia region).

Institutions and Partnerships. NCARE—various departments; Drought Monitoring Unit, Ministry of Agriculture; Jordan Meteorological Department; University of Jordan; Jordan Institute of Science and Technology; Badia Research and Development Center; Royal Geographic Center; rural communities.

Improve Livestock and Rangeland Systems

Objectives and Proposed Activities. Livestock remains a key component of local agricultural systems in low-moisture upland areas of the Jordan River Valley, and particularly elsewhere in Jordan outside the Jordan River Valley. The activities proposed here have the overall objectives of improving livestock productivity, especially for small ruminants (for example, sheep and goats), and improving the sustainable management of rangelands under expected climatic changes. Improving livestock productivity can be achieved via proposed activities such as: (1) improving the genetic characteristics of endemic livestock through the development of improved breeds that are more tolerant of dry climate conditions; and (2) improving livestock management practices in light of expected climate changes. These practices include: improving animal nutrition, improved grazing management (rotations, use of legume forages), etc. The sustainable management of rangelands can be enhanced through activities that promote: improved monitoring of rangeland conditions; rangeland rehabilitation; maintenance of biodiversity and rangeland resources that can be used in adapting to climate change. In both cases, increasing awareness of climate change and land-use impacts among farm households and rural communities are key.

Expected Results. (1) development of heat-tolerant breeds; (2) improvement of the sustainability and productivity of rangelands; (3) improvements in livestock production and rural household incomes; and (4) increased awareness of climate change impacts and land-use alternatives among rural communities.

Institutions and Partnerships. Integrated Livestock and Rangeland Department, NCARE; Biodiversity and Medicinal Plants Directorate, NCARE; rural communities.

Build National Capacity for Climate Change Adaptation

Objectives and Proposed Activities. The challenges posed by climate change in the Jordan River Valley and elsewhere in Jordan require substantially improving the capacity of farmers, the government, agricultural researchers and rural communities to deal with anticipated climate changes. Proposed activities to address these capacity-building needs include: improving the technical knowledge base of NCARE staff and the relevance of NCARE's research to deal with climate change problems; building an integrated regional database and capacity to track climate changes in the Jordan River Valley; strengthening collaborations and communications with farmers; and, overall, improving the capacity of national institutions and rural people to deal with climate risk. Development of a National Agricultural Climate Change Strategy is needed, perhaps as a follow-on to the initial steps called for in the Agricultural Sector Strategy (2011–2013); this could stimulate a concerted national effort to involve new political, social, and economic partnerships to address climate change in agriculture. Institutional commitments are needed to ensure the necessary capacity-building, information generation, improved institutional and legal arrangements, technology development and associated funding. Improved systems for climate monitoring and the planning and implementation of climate change mitigation and adaptation response options are needed. Existing mandates and capacities regarding natural resource management are currently scattered among various public institutions; these should be consolidated.

Expected Results. The long-run result would be an improved national capacity— of institutions, farmers, researchers and others—to address climate change in agriculture.

Institutions and Partnerships. National Center for Agricultural Research and Extension; various government ministries; farmers and rural communities; and other institutions related to water management and the environment.

Reduce Risks of Agricultural Pests and Diseases

Objectives and Proposed Activities. Increases in the frequency and severity of crop pest and disease outbreaks, weed infestations, as well as the introduction of new pests and diseases, are likely to occur in the future due to climate changes. Increasing the capacity of farmers to address these growing problems is important in maintaining agricultural productivity and food security. Activities proposed here are: (1) to study the impacts of climate on population dynamics of plant diseases and pests; and (2) to identify and promote suitable management and cultural practices to reduce future infestations, including integrated pest management practices.

Expected Results. Reductions in the increased frequency and severity of crop pests, diseases and weed infestation, and associated improvements of farm households' incomes.

Institutions and Partnerships. Land Protection Research Department, the National Center for Agricultural Research and Extension; farmers; government ministries; other institutions.

Reinforce Early Warning System for Drought

Objectives and Proposed Activities. Jordan has a drought early warning system (DEWS) that was introduced by the Drought Monitoring Unit (DMU) of NCARE following its establishment in 2008. The DMU monitors drought through the identification of drought severity and its geographic distribution over the country using the remotely sensed normalized difference vegetation index (NDVI) plus seasonal rainfall records. NDVI maps are produced every 16 days using MODIS images (depending on their availability). Given the current importance of drought in Jordan and the expected increased frequency and severity of drought with future climatic changes, further attention needs to be paid to drought monitoring alternatives and effectiveness in the country. Three specific activities are proposed here: (1) Analyze different alternatives for remotely sensed indices of drought (vegetation condition index (VCI), temperature condition index (TCI), vegetation health index (VHI), etc.) on a real-time basis, and evaluate the attributes of each as applied to the Jordanian situation; (2) conduct field studies of drought effects and their correlations with different remotely sensed indices; and (3) perform a socio-economic study for selected affected areas to identify the potential economic damages of drought (for example, yields), and how they correlate with different drought indices and indicators (NDVI maps, field vegetation status, rainfall patterns).

Expected Results. Develop more effective methods and indicators to minimize the impacts of drought on people and agriculture in the Jordan River Valley and elsewhere in Jordan. Long-term benefits including increased farm household incomes and agricultural sustainability will result from a better ability to respond to drought.

Institutions and Partnerships. Drought Monitoring Unit (DMU), NCARE; Jordan Meteorological Department; Ministry of Water and Irrigation Water, Soil and Environment; various NCARE departments: Field Crops, Horticulture, Integrated Livestock and Rangeland, Socio Economic Studies. International collaborators of the DMU include the Arab Center for the Study of Arid Zones and Dry Lands (ACSAD) and the World Food Programme.

Reform Land-Use Laws and Implement Sustainable Land-Use

Objectives and Proposed Activities. Land-use in Jordan suffers from many problems, including land fragmentation, conflicts among competing uses, and desertification. Improved land-use planning should be supported by appropriate legislation to balance the demands of multiple users and to encourage land-use changes and improved land management practices. These include: planting according to land suitability, improving cultivation practices (including conservation agriculture), reducing overgrazing, introducing water harvesting techniques, and others, in order to reduce desertification and soil erosion and to overcome the challenges posed by land fragmentation. To support movement in the direction of sustainable land management, activities proposed under this response option include: (1) Reviewing existing laws and propose changes to enhance sustainable land-use practices; (2) improving the mapping of soils, land-use status and desertification risk; and (3) studying and analyzing land-use and land cover, land ownership, parcel size, land fragmentation and land-use suitability, and identifying appropriate soil and water conservation techniques that apply to different biophysical circumstances.

Expected Results. Improved and more sustainable land-use management; improved land-use planning.

Institutions and Partnerships. Ministry of Agriculture; NCARE; Ministry of Water and Irrigation; Ministry of Environment; Ministry of Minicipal Affairs; Ministry of Justice; Department of Land and Surveys, Ministry of Finance; Prime Minister's Office.

Activation of Agricultural Risk Management Fund

Objectives and Proposed Activities. As in many Arab countries, no agricultural insurance system exists in Jordan. The government has sought to establish the groundwork for agricultural insurance, and the regulations have been approved for an Agricultural Risk Management Fund, but it has not started functioning due to lack of financing. The Government, however, acts as a "quasi-insurer" by providing compensation to farmers whose production suffered from drought, frost or flooding (EBRD 2011). This should improve farmers' abilities to respond to climate-related risks. The proposed activity is to identify legislation, regulations, tools, financing and infrastructure needed for the Fund to function effectively, and implement it, if feasible.

Expected Results. Effective functioning of the Agricultural Risk Management Fund.

Institutions and Partnerships. Ministry of Agriculture; Agricultural Risk Management Fund; private insurance companies; and farmers.

Box 3.3

A Comparison of Selected Agricultural Indicators for Jordan and Lebanon

Indicator (unit)	Jordan	Lebanon	Source
Country-General			
GDP (current US$) [2010]	27,573,536,000	39,006,223,284	WB
Population, total [2010]	6,047,000	4,227,000	WB
Income level	Upper middle	Upper middle	WB
GNI per capita (in US$) [1999]	1630	3700	—
HDI Rank (out of 162 countries) [1999]	88	65	—
Value			
Agriculture's contribution to GDP (%) [2010]	3	6	WB
Average annual growth (%) [2010]	6.6	3.5	WB
Agriculture, value added (% of GDP) [2010]	3	6	WB
Agriculture, value added (constant 2000 US$) [2008]	267,421,312	1,106,254,350	WB
Food security and Food Aid			
Official Development Assistance to agriculture (2004 US$ millions)	3	5.6	WB
Official Development Assistance to agriculture (% total to country)	0.3	3	WB
Cereals food aid 1,000 tons [2003–2005]	98.3	11	WB
Self-sufficiency ratio (%)-cereals [2007]	2.07	18.05	UNDP
Self-sufficiency ratio (%)-meat [2007]	68.14	86.95	UNDP
Self-sufficiency ratio (%)-milk and dairy [2007]	50.89	31.18	UNDP
Self-sufficiency ratio (%)-fruits [2007]	89.13	122.98	UNDP
Self-sufficiency ratio (%)-vegetables [2007]	203.71	95.3	UNDP
Rural population			
Rural population (% of total population) [2010]	22	13	WB
Rural population density (pop per sq. km of arable land) [2005]	521	289	WB
Rural population growth (annual %) [2008]	3	0.1	WB
Population economically active in agriculture (% total active population) [2005]	9.8	2.6	FAO
Share of women in agricultural labor force (%) [2003–2005]	70.1	40	FAO
Agriculture value added per worker (constant 2000 US$) [2010]	3,401	41,013	WB
Rural population below national poverty line (% of total population)	12	27	IFAD
Rural population below national poverty line (% of total poor)	12	25	IFAD
Stunting prevalence in children under 5 years (rural)	27.3	—	IFAD
Stunting prevalence in children under 5 years (urban)	15.8	—	IFAD

(box continues on next page)

Box 3.1 A Comparison of Selected Agricultural Indicators for Jordan and Lebanon *(continued)*

Indicator (unit)	Jordan	Lebanon	Source
Land			
Agricultural land (% of land area) [2007]	10.9	67.3	WB
Arable land (hectares per person) [2009]	0.03	0.03	WB
Arable land (hectares per agricultural person) [2003–2005]	3.1	0.5	WB
Gini index	0.78	0.69	WB
Agricultural irrigated land (% of total agricultural land) [2007]	7.5	19.9	WB
Permanent cropland (% of land area) [2009]	0.9	14	WB
Land under cereal production (hectares) [2010]	44,469	64,940	WB
Water			
Renewable internal freshwater resources per capita (m^3) [2005]	129	1,197	WB
Annual fresh water withdrawals for agriculture % total [2002]	75	67	WB
Agricultural production			
Technical efficiency (index from 0 to 1)	0.61	0.88	H&K (IMF)
Total Factor Productivity Growth (%)	0.97	1.31	H&K (IMF)
Cereal yield (kg per hectare) [2010]	1,963	2,740	WB
Crop production index (2004–2006 = 100) [2009]	106	101	WB
Fertilizer consumption (kg/ha of arable land) [2003–2005]	498	96	WB
Food production index (2004–2006 = 100) [2009]	110	104	WB
Livestock production index (2004–2006 = 100) [2009]	115	112	WB

Sources: WB: World Bank (2012) and World Bank (2008), UNDP: UNDP (2008), FAO: FAO (2008), H&K (IMF): Hassine and Kandil (2009). IFAD: IFAD(2003).
Note: — = not available.

Figure B3.3.1 Agriculture Value Added Per Worker (Constant 2000 US$)—2008

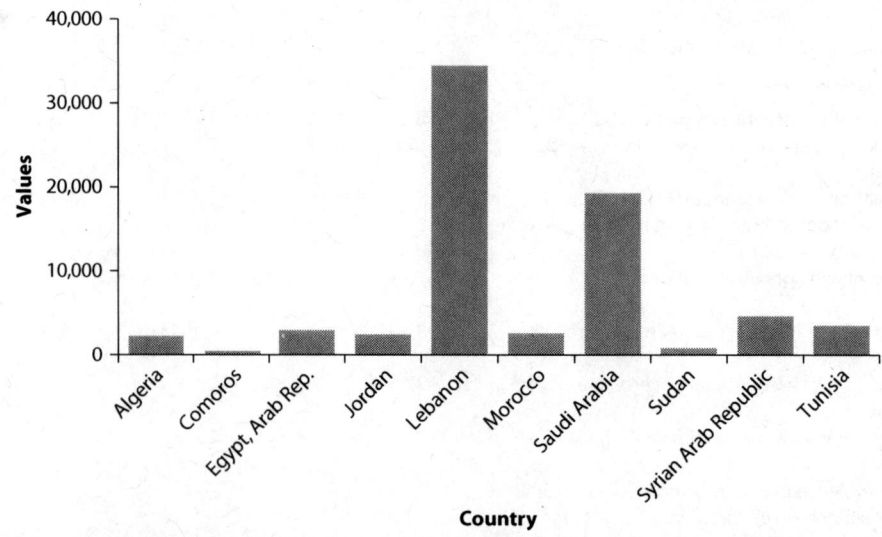

Notes

1. One dunum is approximately 1/10 of one hectare.

2. Fertigation is the application of fertilizers, soil amendments, or other water-soluble products through an irrigation system.

3. Among the impact criteria, the net economic benefits (for example, benefits minus costs; weight = 19.7) and the potential to promote climate adaptation (weight = 19.2) associated with each response option were identified by participants as the top-ranked criteria. The next three impact criteria received roughly equal prioritization by participants: "Other environmental impacts," such as biodiversity preservation and mitigating soil degradation, "Complementarity between public and private sectors," and the "Importance of the activity to the poor and local communities," received proportionate weights of 15.4, 14.2, and 13.0, respectively. The lowest ranked impact criteria were "Time required to achieve positive results," (weight = 10.4) and the "Indirect and spillover effects on other sectors" (weight = 8.6). Among the viability criteria, the highest ranked were the technical viability of the response option (weight = 19.0) and the "Importance of public sector intervention" (weight = 17.2). The other criteria— degree of public support, availability and quality of information needed, compatibility with the national climate change strategy, and level of preparedness to undertake implementation of the response option—were all similarly weighted (weights between 15.5 and 16.5).

4. Moderate Resolution Imaging Spectroradiometer operated by NASA (US National Aeronautics and Space Administration).

References

Abdel-Wali, M. 2010. "Assessing Climatic Changes in the Biotic Environment of the Agricultural System." World Bank Workshop presentation, Amman, Jordan, October.

ACS, 2006. "Compendium statistique national sur les statistiques de l'environnement au Liban 2006." Administration Centrale de la Statistique. Beirut, Lebanon. http://www.cas.gov.lb/images/PDFs/enviromment.pdf.

Adams, R. H. 2001. "Nonfarm Income, Inequality and Poverty in Rural Egypt and Jordan." PRMPO, World Bank, Washington, DC.

Ahmadi, M., E. M. Heravan, S. Y. Sadeghian, M. Mesbah, and F. Darvish. 2011. "Drought Tolerance Variability in S1 Pollinator Lines Developed from a Sugar Beet Open Population." *Euphytica* 178: 339–49.

Al-Bakri, J., A. Suleiman, F. Abdulla, and J. Ayad. 2010. "Potential Impact of Climate Change on Rainfed Agriculture of a Semi-Arid Basin in Jordan." *Physics and Chemistry of the Earth* 35: 125–34.

Al-Jaloudy, M. A. 2006. "Country Pasture/Forage Resource Profiles—Jordan." Food and Agriculture Organization of the United Nations, Rome. http://www.fao.org/ag/agp/agpc/doc/Counprof/PDF%20files/Jordan.pdf.

Al-Naber, G. 2010. "*Jordan River Basin*." World Bank Reducing Vulnerability to Climate Change in Agricultural Systems workshop presentation, Amman, Jordan, October.

Alston, J. M., G. Norton, and P. Pardey. 1995. *Science Under Scarcity: Principles and Practice for Agricultural Research Evaluation and Priority Setting*. Ithaca and London: Cornell University Press.

Amery, H.. 2002. "Irrigation Planning in Lebanon: Challenges and Oportunities." In *Modern and Traditional Irrigation Technologies in the Eastern Mediterranean*, edited by Ö. Mehmet and H. A. Biçak. Ottawa, Canada: International Development Research Center.

Ashwill, M., C. Flora, and J. Flora. 2011. *Building Community Resilience to Climate Change: Testing the Adaptation Coalition Framework in Latin America*. Washington, DC: World Bank, November. http://siteresources.worldbank.org/EXTSOCIALDEVELOPMENT/Resources/244362-1232059926563/5747581-1239131985528/Adaptation-Coalition-Framework-Latin-America_web.pdf.

Asmar, F. R.. 2011. "Country Pasture/Forage Resource Profiles—Lebanon." Food and Agriculture Organization of the United Nations, Rome. http://www.fao.org/ag/AGP/AGPC/doc/Counprof/PDF%20files/Lebanon.pdf.

Baltussen, R., and R. Niessen. 2006. "Priority Setting of Health Interventions: The Need for Multi-criteria Decision Analysis." *Cost Effectiveness and Resource Allocation* 4: 14.

Bani Hani, A. 1996. "Determinants of Agricultural Labor Wage in Jordan: A Comparative Study of Expatriate versus Local Labor." *Economic Development and Cultural Change* 44 (2): 405–11.

Bou-Zeid, E., and M. El-Fadel. 2002. "Climate Change and Water Resources in Lebanon and the Middle East." *Journal of Water Resources Planning and Management-ASCE* 128: 343–55.

Breisinger, C., T. van Reheenen, C. Ringler, A. N. Pratt, N. Minot, C. Aragon, B. Yu, O. Ecker, and T. Zhu. 2010. "Food Security and Economic Development in the Middle East and North Africa: Current State and Future Perspectives." IFPRI Discussion Paper 00985, International Food Policy Research Institute, Washington, DC.

CIA (Central Intelligence Agency). 2012. "The World Factbook." Central Intelligence Agency, Washington, DC. https://www.cia.gov/library/publications/the-world-factbook/geos/le.html.

Choueiry, E., and C. Hobaika. 2010. *"The Impact of Climate Change on Agricultural Pests and Diseases in Lebanon."* World Bank Reducing Vulnerability to Climate Change in Agricultural Systems workshop presentation, Lebanon, October.

Christensen, J. H., B. Hewitson, A. Busuioc, A. Chen, X. Gao, I. Held, R. Jones, R. K. Kolli, W.-T. Kwon, R. Laprise, V. Magaña Rueda, L. Mearns, C. G. Menéndez, J. Räisänen, A. Rinke, A. Sarr and P. Whetton, 2007: Regional Climate Projections. In: Climate Change 2007: The Physical Science Basis. Contribution of Working Group I to the Fourth Assessment Report of the Intergovernmental Panel on Climate Change [Solomon, S., D. Qin, M. Manning, Z. Chen, M. Marquis, K. B. Averyt, M. Tignor and H. L. Miller (eds.)]. Cambridge University Press, Cambridge, United Kingdom and New York, NY, USA.

Collelo, T., ed. 1987. *Lebanon: A Country Study*. 3rd ed. Federal Research Division, Library of Congress. Washington: GPO for the Library of Congress.

Commandeur, P. 1997. "The DGIS Special Programme on Biotechnology." *Biotechnology and Development Monitor* (31): 6–11.

DOS (Department of Statistics). 2012. "Agricultural Survey of 2007." Department of Statistics, Jordan (accessed April 29, 2012). http://www.dos.gov.jo/agr/agr_e/index.htm.

EAT/USAID (U.S. Agency for International Development/Enabling Agricultural Trade). 2012. "AgBEE Snapshot: Jordan." Enabling Agricultural Trade, U.S. Agency for International Development, Washington, DC.

Embassy, J. 2012. "Media and Communications: Agriculture." Embassy of the Hashemite Kingdom of Jordan, Washington, DC. http://www.jordanembassyus.org/new/jib/factsheets/agriculture.shtml.

Environmental Assessment Institute. 2006. "Risk and Uncertainty in Cost-Benefit Analysis: A Toolbox Paper." Environmental Assessment Institute, Copenhagen, Denmark.

EBRD (European Bank for Reconstruction and Development). 2011. Jordan's Request for Country of Operations Status: Technical Assessment, November. http://www.ebrd.com/downloads/country/technical_assessments/2012-02-13_Jordan_TA.pdf

Evans, J. P. 2009. "21st Century Climate Change in the Middle East." *Climatic Change* 92: 417–32.

Evans, J. P., R. B. Smith, and R. J. Oglesby. 2004. "Middle East Climate simulation and Dominant Precipitation Processes." *International Journal of Climatology* 24: 1671–94.

Falconi, C. 1999. "Methods of Priority Setting in Agricultural Biotechnology Research. Ch. 4." In *Managing Agricultural Biotechnology—Addressing Research Program Needs and Policy Implications*, edited by J. I. Cohen. Wallingford, Oxon: CAB International.

FAO (Food and Agriculture Organization). 1997. "Irrigation in the Near East Region in Figures." Food and Agriculture Organization of the United Nations, Rome. http://www.fao.org/docrep/W4356E/W4356E00.htm.

———. 2007a. "Irrigation in the Near East Region in Figures." Food and Agriculture Organization of the United Nations, Rome (accessed April 27, 2012). http://www.fao.org/docrep/W4356E/W4356E00.htm.

———. 2008. "Irrigation in the Middle East Region in Figures-Aquastat Survey." Food and Agriculture Organization of the United Nations, Rome. http://www.fao.org/nr/water/aquastat/countries/lebanon/index.stm.

———. 2009. "Irrigation in the Middle East Region in Figures-Aquastat Survey." Food and Agriculture Organization of the United Nations, Rome. http://www.fao.org/docrep/012/i0936e/i0936e00.htm.

———. 2010. "Exports: Commodities by Country, Lebanon, 2009." Food and Agriculture Organization of the United Nations, Rome. http://faostat.fao.org/site/342/default.aspx.

———. 2011. "Regional Priority Framework for the Near East." FAO Regional Office for the Near East, Rome, Italy. http://neareast.fao.org/FCKupload/File/RPF-EN.pdf.

Farajalla, N., M. Marktanner, L. Dagher, and P. Zgheib. 2010. "The National Economic, Environment and Development Studies (NEEDS) for Climate Change Project: Final Report." United Nations Framework Convention on Climate Change. http://unfccc.int/cooperation_and_support/financial_mechanism/items/5630.php.

Fleisher, D. H., D. J. Timlin, and V. R. Reddy. 2008. "Interactive Effects of Carbon Dioxide and Water Stress on Potato Canopy Growth and Development." *Agronomy Journal* 100 (3): 711–19. https://www.crops.org/publications/aj/abstracts/100/3/711.

Fufa, F., S. Grand, O. Kafawin, Y. Shakhatreh, and S. Ceccarelli. 2010. "Efficiency of Farmers' Selection in a Participatory Barley Breeding Programme in Jordan." *Plant Breeding* 129: 156–61.

Garg, N., and D. S. Cheema. 2011. "Assessment of Fruit Quality Attributes of Tomato Hybrids Involving Ripening Mutants under High Temperature Conditions." *Scientia Horticulturae* 131: 29–38.

Haddad, M.. 2009. "Overview of Available Water Resources and Uses in the Jordan Rift Valley." In *The Water of the Jordan Valley: Scarcity and Deterioration of Groundwater and its Impact on the Regional Development*, edited by H. Hötzl, P. Möller and E. Rosenthal. Berlin, Heidelberg: Springer-Verlag.

Hamadeh, S. K., R. Zurayk, F. El-Awar, S. Talhouk, D. Abi Ghanem, and M. Abi-Said. 1999. "Farming System Analysis of Drylands Agriculture in Lebanon: An Analysis of Sustainability." *Journal of Sustainable Agriculture* 15 (2–3): 33–43.

Hartwich, F.. 1999. "Weighting of Agricultural Research Results: Strength and Limitations of the Analytic Hierarchy Process." Discussion Paper 09–99, University of Hohenheim, Stuttgart, Germany. http://entwicklungspolitik.uni-hohenheim.de/uploads/media/DP_09_1999_Hartwich.pdf.

Harvell, C. D., C. E. Mitchell, J. R. Ward, S. Altizer, A. P. Dobson, R. S. Ostfeld, and M. D. Samuel. 2002. "Climate Warming and Disease Risk for Terrestrial and Marine Biota." *Science* 296: 2158–62.

Hassine, N., and M. Kandil. 2009. "Trade Liberalization, Agricultural Productivity and Poverty in the Mediterranean Region." *European Review of Agricultural Economics* 36 (1): 1–29.

Haverkort, A. 2008. "Climate Change: Impact and Opportunities for Potato. Proceedings of the International Year of the Potato: Improving International Potato Production." *Dundee* (August).

Houri, A., and El. Jeblawi. 2007. "Water Quality Assessment of Lebanese Coastal Rivers During Dry Season and Pollution Load into the Mediterranean Sea." *Journal of Water and Health* 5 (4): 615–23. http://www.iwaponline.com/jwh/005/0615/0050615.pdf.

IPCC (Intergovernmental Panel on Climate Change). 2001. *Impacts, Adaptation and Vulnerability, Contribution of Working Group II to Third Assessment Report of the Intergovernmental Panel on Climate Change.* Cambridge, UK: Cambridge University Press.

Institute of Medicine. 1986. *New Vaccine Development: Establishing Priorities, Vol. 2. Committee on Issues and Priorities for New Vaccine Development.* Washington, DC: National Academies Press.

IFAD (International Fund for Agricultural Development). 2002. "329-JO: Income Diversification Project." IFAD Evaluation Knowledge System, Rome.

———. 2003. "Assessment of Rural Poverty-Near East and North Africa." IFAD Evaluation Knowledge System, Rome. http://www.ifad.org/poverty/region/pn/nena.pdf.

IMF (International Monetary Fund). 2010. "Lebanon: Real GDP Growth Analysis." Resident Representative Office in Lebanon, July. http://www.imf.org/external/country/LBN/rr/2010/070110.pdf.

International Water Management Institute. 2012. "Jordan River Basin: Short Profile." Jordan. http://www.iwmi.cgiar.org/Assessment/FILES/word/ProjectDocuments/BasinFactSheets/Jordan%20Basin%20short%20profile.pdf.

Jabarin, A. S. 1997. "Some Expect Impacts of the Peace Treaty on Horticultural Production in the Jordan Valley (Jordan)." *Journal of Economic Cooperation among Islamic Countries* 18: 143–53.

Janssen, W., and A. Kissi. 1997. "Planning and Priority Setting for Regional Research: A Practical Approach to Combine Natural Resource Management and Productivity Concerns." Research Management Guidelines No. 4, International Service for National Agricultural Research, The Hague.

Johanson, L.. 2005. "English Translation of the Agriculture Strategy Developed by the Ministry of Agriculture of Jordan: Final Report." Office of Economic Opportunities USAID/Jordan.

Joumaa, I.. 2010. *"Reducing Vulnerability to Climate Change in Agricultural Systems."* World Bank Reducing Vulnerability to Climate Change in Agricultural Systems workshop presentation, Lebanon, October.

Karam, F., and K. Karaa. 1999. "Recent Trends Towards Developing a Sustainable Irrigated Agriculture in the Bekaa Valley of Lebanon." In Proceedings of the Annual Meeting of the Mediterranean Network on Collective Irrigation Systems, CIHEAM Malta, November 3–6. http://ressources.ciheam.org/om/pdf/b31/010020077.pdf.

Khresat, S. A. 2010. "Assessing Climatic Changes in the Biotic Environment of the Agricultural System." World Bank Reducing Vulnerability to Climate Change in Agricultural Systems workshop presentation, Amman, Jordan, October.

Kuch, P. J., and S. Gigli. 2007. "Economic Approaches to Climate Change Adaptation and their Role in Project Prioritisation and Appraisal." Deutsche Gesellschaft für Technische Zusammenarbeit, Climate Protection Programme, Eschborn, Germany. http://www.additiv.li/ref/gtz_ada_eco.pdf.

Laithy, H., K. Abu-Ismail, and K. Hamdan. 2008. *"Poverty, Growth and Income Distribution in Lebanon."* Country study published by the International Poverty Center (13), Brasilia DF, Brazil.

Leavitt, S. W., S. B. Idso, B. A. Kimball, J. M. Burns, A. Sinha, and L. Stott. 2003. "The Effect of Long-term Atmospheric CO_2 enrichment on the Intrinsic Water-use Efficiency of Sour Orange Trees." *Chemosphere* 50: 217–22.

Manderscheid, R., A. Pacholski, and H.-J. Weigel. 2010. "Effect of Free Air Carbon Dioxide Enrichment Combined with Two Nitrogen Levels on Growth, Yield and Yield Quality of Sugar Beet: Evidence for a Sink Limitation of Beet Growth Under Elevated CO_2." *European Journal of Agronomy* 32: 228–39.

Manicad, G.. 1997. "Priority Setting in Agricultural Research: A Brief Conceptual Background." *Biotechnology and Development Monitor* (31): 2–6.

Mehmet, Ö., and H. A. Biçak, eds. 2002. *Modern and Traditional Irrigation Technologies in the Eastern Mediterranean*. Ottawa, Canada: International Development Research Center.

Ministry of Agriculture (Lebanon) and FAO. 2000. "Résultats globaux du Recensement agricole." Ministère de l'Agriculture, FAO, Projet 'Assistance au recensement agricole'. 122pp.

———. 2007. "Agricultural Statistics, Lebanon 2005." Ministry of Agriculture. Directorate of Studies and Coordination, Beirut, Lebanon.

———. 2008. "Lebanon State of the Environment Report." Ministry of Environment/LEDO, Lebanon.

———. 2011. "Second National Communication to the UNFCCC." February. http://maindb.unfccc.int/library/view_pdf.pl?url=http://unfccc.int/resource/docs/natc/lbnnc2.pdf.

Ministry of Environment (Lebanon)/LEDO/ ECODIT. 2001. "Lebanon State of the Environment Report." Ministry of Environment/Lebanese Environment and Development Observatory and ECODIT Liban, Beirut, Lebanon.

Molle, F., J. P. Venot, and Y. Hassan. 2008. "Irrigation in the Jordan Valley: Are Water Pricing Policies Overly Optimistic?" *Agricultural Water Management* 95: 427–38.

Moriondo, M., F. M. Stefanini, and M. Bindi. 2008. "Reproduction of Olive Tree Habitat Suitability for Global Change Impact Assessment." *Ecological Modelling* 218: 95–109.

Ober, E. S., and A. Rajabi. 2010. "Abiotic Stress in Sugar Beet." *Sugar Technology* 12 (3–4): 294–98.

Peet, M. M., D. H. Willits, and R. Gardner. 1997. "Response of Ovule Development and Post-pollen Production Processes in Male-sterile Tomatoes to Chronic, Sub-acute High Temperature Stress." *Journal of Experimental Botany* 48: 101–11.

Pizer, W. 2005. "Climate Policy Design Under Uncertainty." Discussion Paper 05–44. Resources for the Future, Washington, DC. October.

Qi, A., and K. W. Jaggard. 2006. "Partitioning Climate Drought into Effects of Water Stress and Hot Temperature in the UK." *Sugar Industry/Zuckerindustrie* 131 (6): 412–15.

Qumair, F. 1998. "Pollution Problems and Surface Water Losses." *Abaad* 7: 14–19.

Saba, M., G. Al-Naber, and Y. Mohawesh. 2010. "Analysis of Jordan's Vegetation Cover Dynamics using MODIS/NDVI from 2000–2009." In Food Security and Climate Change in Dry Areas. Conference Proceedings, International Center for Agricultural Research in Dry Areas, Aleppo, Syria.

Samuels, R., S. Krichak, P. Alpert, A. Rimmer, and A. Hartmann. 2010. "Climate Change Impacts on Jordan River Flow: Downscaling Application from a Regional Climate Model." *Journal of Hydrometeorology* 11 (4): 860.

Saoub, H. 2010. "Assessing Impact of Climate Change on Potential and Existing Agricultural Activities." Presentation at the World Bank Workshop Reducing Vulnerability to Climate Change in Agricultural Systems, NCARE and World Bank, Amman, Jordan, October 6–7.

Schwarz, D., Y. Rouphael, G. Colla, and J. H. Venema. 2010. "Grafting as a Tool to Improve Tolerance of Vegetables to Abiotic Stresses: Thermal Stress, Water Stress and Organic Pollutants." *Scientia Horticultuerae* 127: 162–71.

Shaban, A. 2009. "Indicators and Aspects of Hydrological Drought in Lebanon." *Water Resources Management* 23: 1875–91.

Shatanawi, M., A. Fardous, N. Mazahrih, and M. Duqqah. 2005. "Irrigation Systems Performance in Jordan." In *Irrigation Systems Performance*, edited by N. Lamaddalena, F. Lebdi, M. Todorovic, and C. Bogliotti, 123–31. Bari: CIHEAM-IAMB. (Options Méditerranéennes: Série B. Etudes et Recherches; n. 52). 2. WASAMED Workshop: Irrigation Systems Performance, 2004/06/24-28, Hammamet, Tunisia.

Shatanawi, M., J. Al–Bakri, and A. A. Suleiman. 2007. "Lemon Evapotranspiration and Yield Under Water Deficit in Jordan Valley." In *Water Saving in Mediterranean Agriculture and Future Research Needs [Vol. 1]*, edited by N. Lamaddalena, C. Bogliotti, M. Todorovic, and A. Scardigno, 63–71. Bari: CIHEAM-IAMB.

Sheehan, S.. 2008. *Cultures of the World - Lebanon*. 2nd ed. Tarrytown, NY: Marshall-Cavendish.

Teviotdale, B. L., and W. H. Krueger. 2004. "Effects of Timing on Copper Sprays, Defoliation, Rainfall, and Inoculum Concentration on Incidence of Olive Knot Disease." *Plant Disease* 88 (2): 131–35.

Thiombiano, L., and W. Andriesse. 1998. "Research Priority Setting by a Stepped Agro-ecological Approach: Case Study for the Sahel of Burkina Faso." *Netherlands Journal of Agricultural Science* 46: 5–14.

Valerio, M., M. B. Tomecek, S. Lovelli, and L. H. Ziska. 2011. "Quantifying the Effect of Drought on Carbon Dioxide-induced Changes in Competition between a C3 Crop (Tomato) and a C4 Week (Amaranthus retroflexus)." *Weed Research* 51: 591–600.

Venot, J. P. 2003. *Farming Systems in the Jordan River Basin in Jordan: Agronomical and Economic Description.* Paris, France: Institut National d'Agronomie de Paris-Grignon.

Verdeil, É., G. Faour, and S. Velut. 2007. "Chapitre 5: L'économie." In *Atlas du Liban,* 116–35. Beirut, Lebanon: Institut français du Proche-Orient/CNRS Liban. http://ifpo. revues.org/420.

Wilby, R. L. 2010. "Climate Change Projections and Downscaling for Jordan, Lebanon and Syria: Draft Synthesis Report." World Bank, MENA Region, September.

World Bank. 2009. *Building Response Strategies to Climate Change in Agricultural Systems in Latin America.* Washington, DC: Latin American and the Caribbean Region.

———. 2012. "Data, Indicators." World Bank, Washington, DC (accessed April 25, 2012). http://data.worldbank.org/country.

Zeid, M. A. 2005. "Baseline Study for Table Grapes and Potatoes in Lebanon." Regional Integrated Pest Management Program in the Near East. GTFS/REM/070/ITA, May-November, Damascus, Syria.

Zeid, M. A. 2007. "Baseline Study for Apples and Olives in Lebanon." Regional Integrated Pest Management Program in the Near East. GTFS/REM/070/ITA.

Zhang, J., X. Jiang, T. Li, and J. Xu. 2011. "Effect of Moderately high Temperature Stress and Recovery on the Photosynthetic Characteristics of Tomato (Lycopersicon esculentum L.)." *Journal of Horticultural Science and Biotechnology* 86: 534–38.

CHAPTER 4

Conclusions and Policy Options

Photograph by Dorte Verner

The objectives of this study were threefold: (1) to improve the understanding of climate change impacts on agricultural livelihoods and rural communities in selected regions of Jordan and Lebanon; (2) to engage local stakeholders—including community members, farmers, local experts, and local and national government representatives—in crafting and prioritizing local and regional adaptation strategies and response options that address agricultural sector impacts of anticipated climate changes in the two regions; and (3) to develop local and regional climate change action plans that propose specific recommendations for

107

investment strategies in agricultural research and local agricultural systems. Completion of the priority-setting workshops and development of the regional action plans leads to a number of conclusions, both related to the substantive outcomes of the priority-setting focus of this project, as well as to the context in which these outcomes were generated. In light of these conclusions, several practical recommendations, or policy options, are then made for Jordan and Lebanon's policy makers.

Conclusions

In Jordan and Lebanon, there was a high degree of commonality in terms of the prioritized response options from each country's draft Action Plans. As a consequence, this strengthens the argument that these are urgent actions to be taken, generally, for both countries (see table 4.1). In both Action Plans, addressing water and irrigation-related constraints ranked at the top in terms of priorities. In Lebanon, the two top-ranked response options were explicitly related to irrigation and water management: promoting the adoption of new irrigation technologies through demonstration projects related to drip irrigation—with the potential to greatly economize on water use—and fertigation technologies; and establishing a pilot program for the construction of small- and medium-scale water harvesting reservoirs to provide water storage and supplementary irrigation. The top-ranked response option in Jordan was increasing farm production and efficiency, but it should also be emphasized that many of the proposed activities under this response option relate to improving the efficiency of water use. These activities include: avoiding agricultural expansion into fragile rainfed lands; introducing drought-tolerant crop varieties; identifying alternative cropping patterns that recognize water-related constraints; and promoting conservation agriculture in dry areas. The second-ranked response option in Jordan, increasing water efficiency, was explicitly related to water management. This option encompasses a variety of approaches to improve on-farm water use efficiency and the integrated management of water resources. This includes rainfall harvesting, assessing the feasibility of using treated wastewater and brackish water for irrigation, and developing a system for strict monitoring of groundwater to prevent overexploitation. As expected, water-related constraints were the dominant concern of most local stakeholders participating in the project workshops and were generally the most highly prioritized.

There were other priority response options shared across Lebanon and Jordan, beyond proposed water-related interventions (see table 4.1). These included the development of crop varieties that are tolerant of drought, heat, and other expected climatic changes (in Jordan, subsumed under "Increasing farm production and efficiency"); a focus on integrated pest management (IPM); and the development of improved local capacity to adapt to climate related impacts on agriculture. Improved local capacity was generally interpreted

Table 4.1 Complementary Response Options in Action Plans: Lebanon and Jordan

Umbrella response option	Lebanon	Jordan
Water and irrigation	• Develop drip irrigation and fertigation demonstration projects, including extension activities. (1) • Pilot project to construct small- and medium-scale water harvesting reservoirs (joint with potato seed production). (2)	• Improve farm production systems and productivity—new crop varieties, alternative cropping practices, conservation of genetic diversity, and many water related activities. (1) • Improve on-farm water use efficiency and integrated water resources management with rainfall harvesting, irrigation practices, alternative water sources, groundwater monitoring, etc. (2)
Pest and disease management	• Promote integrated production management of pest, disease and plant physiology disorders for selected high-value fruits, including identification and development of best management practices (BMPs). (3–4)	• Reduce the increasing risks of agricultural pest and disease outbreaks (including introduction of new pests and diseases) outbreaks due to climate change. (6)
Develop new crop varieties/drought resistant livestock	• Public-private partnership to develop certified plant materials (rootstocks, etc.), especially for Prunus varieties, that respond to climatic changes—drought tolerant, resistant to pests and disease, etc. (4) • Maintain genetic diversity of dominant wild species and local varieties adapted to climatic change for selected key crops (wheat, barley, Prunus varieties, fig, caper). (6)	• Improve farm production systems and productivity—new crop varieties, etc. (1) • Improve livestock and rangeland systems—improve livestock genetics, improve animal productivity and sustainable management practices, etc. (3)
Improved capacity	• Enhance capacity building—for farmers and government staff—to deal with climate change adaptations, and including developing rural development strategies to deal with climate change, monitoring and database systems, and creating a national climate change authority. (5)	• Building national capacity in climate adaptation and mitigation, including greater coordination among government institutions. (5) • Develop and reinforce early warning system capacity to deal with drought. (7) • Consider introducing an agricultural insurance system, or "agricultural risk management fund," to improve farmers' abilities to respond to climate-related risks. (9)
Other		• Reform land-use laws and implement sustainable land use, to deal with land fragmentation, desertification and land use conflict. (8)

Source: World Bank data.
Note: The numbers in parentheses indicate the ranking received in the workshop.

broadly to incorporate different types of capacity building among different stakeholders. Some examples of these different types of capacity building include: improved understanding of climate change, information provision and training for farmers; an enhanced general recognition of climate change problems among the general public; and importantly—since this need is often under-recognized in national research systems—improving the institutional capacity of researchers and the research system itself to deal with climate change-oriented problems.

In Jordan, the response options had a stronger policy orientation than in Lebanon. In addition to building capacity, these included: the development of a national climate change strategy; implementation of new land-use laws to foster land-use changes; and implementation of the already approved Agricultural Risk Management Fund (agricultural insurance scheme). The response options in Lebanon assumed more of a research orientation, including research-based initiatives on irrigation technology, crop pests and diseases, and crop varietal development and associated research in applied crop genetics. As expected, many of the response options in Lebanon involved a key role for the Lebanese Agricultural Research Institute (LARI), while in Jordan, the National Center for Agricultural Research and Extension (NCARE) figures prominently in the proposed initiatives.

Overall the Action Plan's priority response options are highly consistent with, and reinforce the importance of, strategic priorities identified in other research. For example, the World Bank's Middle East and North Africa Region (World Bank 2012) has identified three broad areas for strategic partnerships between the Bank and its counterparts to address challenges related to improved climate change adaptation. These include infrastructure investment, knowledge strengthening and policy reform. Most of the abovementioned priority response options fall generally within these categories. Two of the UN Food and Agriculture Organization's regional priorities for responding to climate change impacts in the Middle East region include "improving national and regional capacities to cope with adverse impacts of climate change," and "identifying practices for adaptation and mitigation of climate change impacts," (FAO 2011). Both of these priority areas are directly addressed by the response options related to capacity building. For Lebanon specifically, a recent World Bank review (Lampietti 2010) highlights three challenges facing Lebanese agriculture—infrastructure (irrigation, etc.), water management, and urbanization—and identifies a number of specific strategies to address these challenges; these too are addressed by many of the proposed response options. Finally, the World Bank's new Flagship Report (Verner 2012) on climate change in Arab countries identifies a number of strategies and investments to address climate change in the region. These include: technological innovations; institutional strengthening; improved research tools; farm income diversification; and policy reforms. Each of these was addressed in the specific response options identified and prioritized by local stakeholders in this study.

The priority-setting methodology followed in this study proved to be a practical and viable approach on several levels. The approach incorporates available regional climate projections and thus grounds proposed response options in science-based data and information. In addition, the methodology provides a practical and transparent way to involve local stakeholders in the identification of response options that address climate adaptation needs in agriculture. It allowed for a prioritization of those options under conditions of limited resources. It also led to the drafting of action plans that could be easily communicable at the policy-making level. This bottom-up approach is centered around the input of local stakeholders from the outset, and thus assures that the response options that are prioritized address local needs as viewed by farmers, researchers, extensionists, and others involved at the field-level. There were no discernible problems in eliciting the input of farmers and other local stakeholders on the subject of climate change. Farmers' yields and incomes are directly tied to the natural resource base on which they depend, so they are acutely aware of changes in that resource base—particularly regarding often-limiting water resources—and were eager to share their views and opinions in the workshops organized for this study. Thus, this methodology proved relatively easy, given the workshop-based focus, to engage stakeholders on the subject of climate change in agriculture and related response options.

In general, the development of the action plans served their function as a necessary first step. However, in order to successfully achieve needed investments, interventions, and policies that can serve to locally address climate change impacts on agriculture, this information must be shared with policy makers. Furthermore, these policy makers must be willing to act. In Jordan and Lebanon, the success of this varied. At time of publication, the country team in Jordan had yet to be granted a meeting to deliver the action plan to the country's Minister of Agriculture. A major reason for this is the recent reorganization of the King's Cabinet and high levels of uncertainty over whether or not the current ministers will continue in their posts. In Lebanon, by contrast, the action plan was presented directly to the Minister of Agriculture, who agreed to implement some or all of the priorities. Many details are to still be decided related to the structure, scope and budget of a proposed intervention, but the potential use of this methodology has proved its worth.

In comparing the application of the multi-criteria priority-setting methodology in this project to a similar project in Latin America (World Bank 2009), a couple of additional benefits may be noted. First, execution of the four steps in both regions confirms that the methodology is indeed relatively easy to understand, transparent, and provides an efficient and inexpensive way to garner detailed input from local stakeholders. Second, the workshop setting provides a mechanism for diverse stakeholders to share their views, discuss alternative beliefs about climate change and its local impacts, and evaluate the pros and cons of alternative response options.

The methodology was successful in breaking down the collection of information into a series of "manageable parts." The sequence of several steps in the priority-setting process, which are built around a series of two workshops and a final decision meeting, permits this breakdown. As noted elsewhere (World Bank 2009), this has several advantages. First, it facilitates the distinction between—and reduces confusion among—the identification of response options versus evaluation criteria. By explicitly identifying, assessing and weighting the evaluation criteria, it becomes easier to distinguish between "decision options" and "criteria" for evaluation that sometimes become conflated in participants' minds. Second, by focusing sequentially on four steps: (1) providing information (on climate change and its effects), (2) identifying response options, (3) prioritizing them, and (4) drafting an action plan—the debate and potential contentiousness surrounding steps 3 and 4 is significantly reduced. If the first workshops had begun with a discussion of needed public interventions and policy options to address climate change, the resistance to developing consensus around specific recommendations would likely have been insurmountable (because of grandstanding, promoting of preconceived agendas, etc.). However, by initially focusing on conveying factual information and data on climate change and its observed impacts in the region, it proved possible to develop a common understanding among stakeholders regarding the nature of the underlying problems. The involvement of scientists in the first workshops in providing information on climate changes and their effects helped reduce the potential for conflicting views. This was because most of the scientists and researchers focused on dispassionate presentations of changes, causes, and effects. Importantly, participants did not noticeably interpret this information as biased.

This methodology could be enhanced by mechanisms to improve representation of the most vulnerable groups in the workshops. As with scoring-type methods in general, there is no question that the selection of the workshop participants plays a major role in the outcomes of the study, especially with regard to the weighting of evaluation criteria and the final prioritization of response outcomes. This is a well-known limitation of these methods, but one that is inherent to this approach. An effort was made by the country teams in both Lebanon and Jordan to invite a very diverse set of participants to the workshops in each country to assure that a diversity of views was represented. Given the focus on climate change in agriculture, farmers' views are particularly important since they deal with the daily effects of climate change and represent the "first line of defense" in developing adaptation mechanisms. Thus a particular effort was made to invite and involve farmers in all the workshops, and this effort was generally successful. Many of the farmers involved in the workshops were active and vocal participants. However, the country teams were less successful in assuring the participation of sub-groups of farmers (women farmers, landless farmers, migrant farmers, etc.). This is a limitation because some of these groups are more vulnerable to climate change than others (Ashwill, Flora, and Flora 2011). This shortcoming can easily be rectified by including mechanisms to enhance the diversity of participation, including that of vulnerable groups (World Bank 2011a, 2011b). Despite this, it is important to acknowledge that the workshops had good representation from government ministries and offices, researchers, academics, representatives of nongovernmental organizations (NGOs) and international institutions, journalists and others. The workshops in Lebanon had a relatively high representation of agricultural researchers, thus it is perhaps not too surprising that agricultural research initiatives figured prominently among the prioritized response options.

This methodology is reliant on using workshops to generate knowledge. In addition to possible workshop fatigue,[1] the use of stakeholder workshops as a vehicle to generate response options may prove frustrating to participants if subsequently there is a lack of action in moving forward with the recommendations that are proposed; this was affirmed by some local researchers in Jordan. This is certainly a risk with this methodology since further action is dependent on the feasibility, desire and capability of policy makers to move forward. The approach taken in this study does not automatically generate resources or interest in these initiatives and therefore follow-up actions must be advocated for. Regardless, as a knowledge-generating tool and a mechanism for identifying steps for further interventions, the methodology was highly successful in this case, and at least for Lebanon, is likely to lead to action.

The methodology has a very specific purpose and is not a substitute for necessary scientific and economic analyses. As indicated above, the priority-setting methodology is best at identifying response alternatives from stakeholders, getting stakeholders to evaluate their different strengths and limitations, and generating a prioritization of these alternatives. However, it is not a replacement for cost-benefit analyses or other necessary and detailed technical, economic and institutional assessments of proposed investments and interventions. This is particularly the case with regard to potential major public investments in such areas as irrigation infrastructure. Most of the response options outlined in the action plans did include tentative budgets to initiate activities, at least on a small-scale or pilot basis. However, in most cases, these were only rough estimates by the country teams and did not include a high level of detail, therefore they are excluded from this report.

Policy Options

Agricultural intensification strategies should be implemented in both countries. There are several central challenges facing the food and agricultural sector in Lebanon, Jordan and elsewhere in the Middle East region. These include food security, rural poverty, the critical role of water-related constraints, urbanization and the resulting loss of farmland, and the vulnerability of rural populations to climate change and price volatility. A common thread to addressing these challenges, not only in the Middle East but elsewhere, is the central importance of successful agricultural intensification strategies (Lee et al. 2001; Vosti and Reardon 1997). Agricultural intensification can be defined as cultivating land to achieve the maximum output of crops. The purpose of these strategies is to increase the productivity and income-generating potential of agriculture on an existing or reduced land base. Numerous investments and interventions have been identified in this report to improve agricultural intensification in spite of the challenges presented by climate change. These include: measures to increase the productivity of high-value products, especially fruits and vegetables; improved irrigation and water management; public investments in agricultural research and development; private investments in food marketing and distribution; and a variety of institutional and policy changes to provide a more enabling environment. Each of the activities prioritized in the action plans represent strategies to promote agricultural intensification.

Jordan and Lebanon should focus on improving the production and productivity of value-added agriculture, particularly that of fruits and vegetables. Value-added agriculture can be generally defined as the processing or manufacturing of an agricultural product to enhance its value. An example would be producing wine from grapes. Such value-added strategies in agriculture meet a number of the criteria critical to development in middle-income countries (Cowan 2003; Meijerink and Roza 2007). These criteria include: a high potential for growth in consumer demand; the proximity of both domestic and export markets; high returns per unit of land (particularly important for small landholders); and high levels of diversification both in terms of production and consumption (by contributing to food security through dietary diversification). In terms of climate adaptation, particularly in water-scarce environments of countries like Lebanon and Jordan, value-added agriculture takes on new importance in terms of its economical use of water inputs, its potential to be successful in areas experiencing urban growth and farmland loss (such as in Bekaa Valley), and taking advantage of the local research base.

No- and low-regret adaptation strategies should be pursued in both Jordan and Lebanon. As is evident in chapter 2 of this report, downscaled climate projections for Lebanon and Jordan demonstrate potentially severe impacts from climate change throughout the twenty-first century. These include higher forecasted temperatures (1.3–2.3°C by the 2050s and 1.9–4.0°C by the 2080s); lower precipitation (reductions of 8–29 percent by the 2050s and 14–51 percent by the 2080s), especially in the Bekaa Valley; longer dry seasons; and increasing regional water deficits. These trends are already underway and are expected to be exacerbated in the future. Yet, despite these projections it is still highly uncertain how these changes will impact humans. Will there be more floods? If so, where? Will they lead to increased competition over dwindling resources, migration or social conflict? It is because of this uncertainty that it is important to implement strategies that will have net positive social benefits regardless of climate impacts. Such a no-regrets approach will assure that maladaptive strategies (adaptation strategies that lead to negative outcomes) are not enacted. In this light, no-regret and low-regret climate adaptation options and policies that generate high direct or indirect benefits currently, even in the face of uncertainty regarding future climate impacts, makes sense. The Action Plans from both Jordan and Lebanon include many no- or low-regrets options related to water. These include: improving irrigation and water delivery infrastructure; research on new water management technologies and crop varieties; improving water use efficiency through management practices; improved climate monitoring and early warning systems; and a variety of institutional changes that better enable farms and rural households to respond to the changing environment. Since water scarcity is a major issue even without climate change, policy makers will have no regrets about improving water use efficiency.

Jordan and Lebanon should work to improve climate projection information. In the short- and medium-term, the collection and monitoring of climate data could be improved by expanding the number of weather stations, and by collaborating with other countries in the region to improve the coverage and comparability of data. This effort should be combined with a push to link climate data with impact analyses by making climate data available to policy makers and researchers. Some efforts in this direction have already begun. For example, Lebanon is part of the European Climate Assessment and Dataset (ECA&D) project. This aims to combine the collation of a daily series of observations at meteorological stations with quality control, analyze extremes, and to disseminate both the daily data and the analysis results. This effort to improve climate projection information is gradually being extended across the Middle East.

The accessibility of climate data should be improved in both countries. Several actions can be taken to enhance this accessibility. These include: digitalizing data collected in the past that was stored in formats that can be damaged or difficult to access, and encouraging civil authorities to take responsibility for sharing and making the data available to users. This can be especially important when, for example, meteorological services are under the purview of a Ministry of Defense. Many countries have websites with such data for public use. Still, for security reasons, access to current meteorological data is limited, but it is important that older data (for example, one month or one year) at daily or sub-daily temporal resolution is eventually made publicly available. Ideally, information on the availability, conditions for use, and procedures to access data should be provided and regularly updated (Verner 2012). *Policy makers should consistently consider the input of local stakeholders and mechanisms should be in place to assure this.* Local stakeholders are ultimately those whose livelihoods depend most on the success of strategies and policies related to agricultural adaptation. This isso because the impacts of climate change are highly unique to specific localities and, therefore, local people have the greatest familiarity with the on-the-ground realities of their social and agricultural ecosystems (World Bank 2009). Thus, the recommendations and priorities expressed by local stakeholders are particularly important when considering future investments and options to facilitate climate change adaptation. This is reinforced by the fact that the response options identified and prioritized in the Action Plans echo many of the interventions and strategic investments recommended elsewhere by policy makers, international donors, multilateral organizations and others. These bottom-up recommendations help validate and reinforce the strategies made in other contexts, including policy-driven and top-down strategies. Strategies to promote the continuing involvement of agricultural stakeholders in moving forward are diverse, but include: (1) giving a role to local farmer organizations, watershed councils, and similar institutions in the promotion and execution of climate strategies; (2) promoting the wider use of on-farm trials (and not only experiment stations under "scientific" conditions) by agricultural researchers, for example, in crop varietal development and the development of Integrated Pest Management (IPM) strategies; and (3) assuring representation by local farmers and farmer organizations in regional and national agricultural policy formulation (refer to World Bank 2011a for a framework to help achieve this). To affirm the importance of local stakeholder inputs in policy circles it should be noted that finalized action plans were presented to government officials in Lebanon in May 2012, and the Minister of Agriculture, not only agreed with the conclusions, but is taking steps to implement a strategy based on them to build agricultural resilience to climate change.

New technologies should be utilized in the agricultural sector of both countries, with mechanisms in place for continuous revision and for utilizing new advancements. Many of the response options prioritized by local stakeholders in this project focus on technological solutions to climate adaptation— improved irrigation technologies, water harvesting and storage, the development of drought-tolerant crop varieties, improving technologies for groundwater and climate monitoring, and so forth. These technological solutions are important, and indeed, some – like the development of drought-tolerant varieties – are often viewed as central to effective climate adaptation in agriculture. A yet-to-be published report by Lebanon's Ministry of Environment (Ministry of Environment (Lebanon), UNEP Risoe center, and UNDP 2012) specifically prioritizes a number of technologies related to the agricultural sector. These include: conservation agriculture, risk-coping production systems, selection of adapted varieties and rootstocks, integrated pest management, integrated production and protection for greenhouses, early warning systems that incorporate innovative information and communication technologies, and index insurance. Nonetheless, these are not enough (Huesemann 2003). Technological advances are never permanent; they always have a shelf life. Technological change in agriculture is an ongoing process that is key to achieving continuing productivity improvements, whose impacts can be reinforced and magnified through concurrent attention to improving management. Local stakeholders in both countries understood this and, notwithstanding their prioritization of a number of technological interventions and investments, also highlighted the importance of improved management and capacity building in these technologies. This was indicated through such measures as agricultural extension, dissemination of research results, and building human capacity to deal with future climate changes at all levels—on the part of farmers, government officials, researchers, and others.

The public sector should play a major role in climate change adaptation investments, interventions and policy changes. Ultimately, it is the private decision-maker and resource manager—primarily among farm households—who must make the key decisions regarding resource allocation. These decisions include what crops to plant, how much to produce, and similar decisions. But the prioritized response options suggest a critical role for the public sector in dealing with climate change adaptation in agriculture. That role has previously been summarized as focusing on the "three I's" (World Bank 2009): (1) Investments, such as public investments in irrigation infrastructure that entail significant scale economies and that would not be made otherwise, or in agricultural research that has long been shown to have a high payoff in terms of net benefits; (2) Information, providing better information to farmers and resource managers to enable them to manage resources more efficiently and effectively in light of climate change; and (3) Institutions and policy innovations, which have the potential to change the rules of the game, and provide policy frameworks that create better incentives and regulatory structures for private decision makers, backstop and reinforce technical and research-based solutions. In Lebanon, the response options identified and prioritized by stakeholders included a public-private partnership to develop climate-proof plant materials, the improvement of climate monitoring systems, and the establishment of a national climate change authority. In Jordan, the final priority list of response options included a number of proposed policy options—climate change strategy, land-use laws, agricultural insurance—that could fundamentally affect the overall environment and incentive structure for private decision making over resource use in agriculture.

The public sector has a role in improving the information base available to farmers and farm households. The World Bank's Middle East and North Africa Region's flagship report on climate change (Verner 2012) describes how there is a lack of quality information or data collection activities related to the climate in the Arab region, and even when data are collected they are not consolidated or are unavailable. Yet, individual farmers and rural households, who, as mentioned above, make most of the key decisions regarding resource allocation and management, would benefit greatly from the use of this information. This holds true in the context of climate change adaptation. Much of the information base on which farmers would ideally make their private resource decisions is not available but can be considered a public good—non-excludable and non-rivalrous in demand (Cook and Sachs 1999). As a public good, there is commonly an under-supply of information by the private sector. As a result, sub-optimal resource allocation and management decisions are common, for example, the prevalence of low water use efficiency in irrigation systems—with negative impacts on production, productivity and food security. For these basic reasons, many institutions—including the United Nations Framework

Convention on Climate Change, the United Nations Development Programme, the World Bank, Stockholm Environment Institute, CSIRO-Australia, and others—have prioritized information provision and decision support as a key mechanisms for public sector investments in adaptation (World Bank 2009). Stakeholders in Lebanon and Jordan have prioritized response options that would be strengthened by public sector support. In Lebanon, these include: climate monitoring systems and databases, and technical advice on irrigation management and integrated pest management. In Jordan, these include: improvements in the early warning system for drought, and improvement of the information base on crop management practices and water resource management. The public sector can potentially play a key role in supplying this type of information, thus improving the capacity of farmers and resource managers to address the challenges posed by climate change.

Notes

1. Workshop fatigue refers to the idea that workshops may be overutilized as a development tool. This can happen because practitioners will often organize workshops in order to involve local stakeholders in the decision making process. If overutilized and if action is not forthcoming, local stakeholders may become "fatigued" by these workshops and stop taking them seriously.

References

Ashwill, M., C. Flora, and J. Flora. 2011. "Building Community Resilience to Climate Change: Testing the Adaptation Coalition Framework in Latin America." World Bank, Washington, DC, November. http://siteresources.worldbank.org/EXTSOCIALDEVELOPMENT/Resources/-1232059926563/5747581-1239131985528/Adaptation-Coalition-Framework-Latin-America_web.pdf.

Cook, L. D., and J. Sachs. 1999. "Regional Public Goods in International Assistance." In *Global Public Goods: International Cooperation in the 21st Century*, edited by I. Kaul, I. Grunberg, and I. Stern. Oxford; New York: Oxford University Press for the United Nations Development Programme.

Cowan, T. 2003. *Value-Added Agricultural Enterprises in Rural Development Strategies.* Hauppauge, New York: Nova Science Pub., Inc.

FAO (Food and Agriculture Organization). 2011. "Regional Priority Framework for the Near East." FAO Regional Office for the Near East, Rome, Italy. http://neareast.fao.org/FCKupload/File/RPF-EN.pdf.

Huesemann, M. H. 2003. "The Limits of Technological Solutions to Sustainable Development." *Clean Technologies and Environmental Policy* 5: 21–34.

Lampietti, J. 2010. "The Future of Agriculture in Lebanon under Climate Change," Development Horizons, Middle East Department, World Bank, First/Second Quarter, 2010, pp. 12–14.

Lee, D. R., C. B. Barrett, P. Hazell, and D. Southgate. 2001. "Assessing Tradeoffs and Synergies among Agricultural Intensification, Economic Development and

Environmental Goals: Conclusions and Implications for Policy." In *Tradeoffs or Synergies? Agricultural Intensification, Economic Development and the Environment*, D. R. Lee and C. B. Barrett, eds., Ch. 24. Wallingford, UK: CABI Publishing.

Meijerink, G., and P. Roza. 2007. "The Role of Agriculture in Economic Development." Markets, Chains and Sustainable Development Strategy & Policy Paper 4, Wageningen, Stichting DLO.

Ministry of Environment (Lebanon), UNEP Risoe center, and UNDP. 2012. "Technology Needs Assessment." Ministry of Environment, Beirut, Lebanon. Unpublished report.

Verner, D., ed. 2012. "Adaptation to a Changing Climate in the Arab Countries: A Case for Adaptation Governance and Leadership in Building Climate Resilience." MENA Development Report, World Bank Publications, Washington, DC.

Vosti, S. A., and T. Reardon, eds. 1997. *Sustainability, Growth and Poverty Alleviation: A Policy and Agroecological Perspective*. Baltimore: Johns Hopkins University Press.

World Bank. 2009. *Building Response Strategies to Climate Change in Agricultural Systems in Latin America*. Washington, DC: Latin American and the Caribbean Region.

———. 2011a. "The Adaptation Coalition Framework: Building Community Resilience to Climate Change." Social Development Unit of Latin America and Caribbean Region, World Bank, Washington, DC. http://siteresources.worldbank.org/ EXTSOCIALDEVELOPMENT/Resources/244362-1232059926563/ 5747581-1239131985528/Adaptation-Coalition-Toolkit_Building-Community-Resilience-Climate-Change_web.pdf.

———. 2011b. "Field Guide: Integrating Gender into Climate Change Adaptation and Rural Development." World Bank, Washington, DC, November. http://siteresources. worldbank.org/EXTSOCIALDEVELOPMENT/Resources/244362-1232059926563/5747581-1239131985528/5999762-1321989469080/Bolivia_ Gender_Guide_CC_English.pdf.

———. 2012. "Adaptation to Climate Change in the Middle East and North Africa Region." MENA Region, World Bank, Washington, DC. http://web.worldbank.org/ WBSITE/EXTERNAL/COUNTRIES/MENAEXT/0,contentMDK%3A21596766 ~pagePK%3A146736~piPK%3A146830~theSitePK%3A256299,00.html.

List of Stations Employed in Syria's First National Communication

Station Name	Lat (N)		Lon (E)		Elev	Record	Yrs	DJF (mm)	MAM (mm)	JJA (mm)	SON (mm)	ANNUAL (mm)
Lattakia	35	36	35	46	9	1955–2006	51	430.7	150.9	10.6	170	762.1
Hmmam	35	24	35	56	48	1956–2006	50	456.8	173.9	13.6	164.6	808.8
Safita	34	49	36	8	370	1955–2006	51	590.3	253.5	8.6	232.7	1085.1
Tartous	34	52	35	53	5	1957–2006	49	484.4	165.8	8.7	181.6	840.4
Tel Abiad	36	42	38	57	348	1957–2005	49	138.7	89	1.8	48.1	277.7
Jaraplus	36	49	38	0	351	1955–2006	51	158.2	95.6	4.9	56.6	315.2
Aleppo	36	11	37	14	385	1955–2006	51	162.6	95.4	2.8	58.2	318.9
Atheria	35	22	37	47	460	1974–2006	32	97	56	2.2	30.6	185.7
Meslmieh	36	20	37	14	415	1955–2006	51	164.4	95.5	2.6	62	324.5
Idleb	35	56	36	37	451	1955–2006	51	279.4	131.2	4	83.7	498.2
Hama	35	7	36	24	305	1955–2000	45	182.5	88.3	3.7	58.1	332.5
Salamiyh	35	1	37	2	448	1955–2006	51	154.9	81.4	3.5	51.4	291.2
Al Rastan	34	56	36	44	390	1960–2001	41	204	92	2.3	65.7	363.9
Homs	34	46	36	43	483	1955–2006	51	242.8	97.7	2.8	71	414.3
Damascus Int. Airport	33	26	36	32	610	1955–2006	51	75.2	27.9	0.4	28.5	132
Mezzeh Air. Dam	33	29	36	13	730	1955–1997	42	119.2	46.9	0.4	38.8	205.3
Kharabo	33	30	36	27	620	1955–2006	51	87.9	33.5	0.2	31	152.5
Dara	32	36	36	7	543	1958–2006	48	155	60.1	1	32.6	248.6
Nabek	34	1	36	44	1329	1955–2006	51	53.6	37.5	1.3	26.7	119
Serghayia	33	48	36	10	1409	1962–2005	43	351.4	149.2	0.8	92.6	594.1
Qunetara	35	49	33	8	941	1986–2006	20	393.6	127.6	3.1	93.4	617.8
Sweida	32	44	36	34	1015	1955–2006	51	201.6	89.6	0.3	45.3	336.7
Palmyra	34	33	38	18	400	1955–2006	51	57.4	44	0.4	25.3	127.1
Maskaneh	35	59	37	59	350	1957–1999	42	111.6	65.7	2	38.8	218.1
Deir Ezzor	35	17	40	11	215	1955–2006	51	76	52.9	0.8	23	152.6
Abuo Kamal	34	26	40	55	175	1955–2006	51	62.5	45.3	0.5	19.3	127.5
Raqqa	35	54	38	59	246	1955–2006	51	94.8	66.4	0.9	30.3	192.4
Al Tanf	32	29	38	40	712	1955–2000	45	43	36.2	0.4	25.5	105
Qoumishlie	37	2	41	12	449	1955–2006	51	208.3	149.9	2.6	57.2	418.1
Hassakeh	36	34	40	43	307	1955–2005	51	131.5	99.2	1.2	41.9	273.8

Source: Maweed 2008.

Mann-Kendall Trend Statistics for Annual Precipitation, Mean Maximum, and Minimum Temperatures, Relative Humidity, Evaporation, and Sunshine Duration at Sites Across Jordan

Station	Rain	Mean temp	Max temp	Min temp	Rel Humid	Evap	Sun hours
Baqura	−0.05	1.4	**3.98****	−0.72	−0.37	−0.03	−1.12
Deir Alla	0.31	**2.42***	0.52	**5.15****	1.48	−0.75	**−3.21****
Ghor Safi	−0.36	**3.73****	**3.44****	**3.16****	**3.55****	**−2.27****	**−4.52****
Irbed	−0.44	1.2	−0.09	**3.28****	**2.55****	**−3.53****	−1.95
Al-Rabbah	0.2	1.34	**2.09***	0.1	1.05	**−7.06****	−0.36
Al-Shoubak	**−2.85****	**2.63****	**3.61****	1.32	−1.73	**−2.67****	**−2.36***
Wadi Dhulail	−1.29	**3.96****	**3.64****	**3.69****	1.99	0.03	−0.16
Jordan Univ.	0.67	**4.69****	**2.15***	**4.71****	−1.86	$	$
Madaba	−1.93	1.97	1.38	**3****	**3.02****	$	$
Aqaba A/P	−0.32	0.75	−0.28	**2.65****	**3.51****	**−5.27****	**−2.08***
Ras Muneef	0.71	1.66	**2.06***	0.83	**4.98****	0.11	−0.52
Amman A/P	−1.14	1.03	−1.27	**3.04****	0.94	**−3.75****	**−8.21****
Ruwaished	1.68	**2.74****	1.22	**4.5****	**2.16***	**−3.88****	**−2.69****
Mafraq	−0.83	**2.58****	1.34	**3.49****	1.55	**−2.99****	**−2.67****
Safawi	0.6	**3.74****	**3.53****	1.48	**2.94****	**−3.35****	**−2.76****
Azraq South	**−2.08***	**3.37****	1.45	**3.27****	0.68	€	€
Q.A.I.A/P	−0.22	**5.48****	**5.34****	**4.76****	−0.8	**−3.27****	0.71
Ma'an	−0.22	1.88	0.61	**2.57****	1.27	**−4.79****	**−3.4****
Al-Jafr	−0.05	**4.29****	1.66	**4.12****	**2.45***	€	**−4.3****

Source: Jordan Ministry of Environment, 2009. Jordan's Second National Communication to the United Nations Framework Convention on Climate Change (UNFCCC). Amman, Jordan, 166pp.

Notes: * Indicates significant trend at the 5 percent confidence level; ** Indicates significant trend at the 1 percent confidence level; $ Sign indicates that the meteorological element is not measured at the station; and € Indicates that the time series is too short.

Predictor Variables for Downscaling Daily Mean Temperature (Top Panel) and Daily Precipitation (Bottom Panel) at Test Sites in Jordan, Lebanon, and the Syrian Arab Republic

Sites	Lat./ Lon./ Elev.	Mean (°C)	Temperature predictors	E %
Aleppo, Syrian Arab Republic	36.18N, 37.20E, 384m	17.5	MSLP, USUR, DSUR, F500, H500, V850	68
Amman, Jordan	31.59N, 35.59E, 780m	17.5	MSLP, DSUR, F500, H500, RSUR	73
Beirut, Lebanon	33.82N, 35.43E, 19m	20.1	MSLP, VSUR, Z500, H500, RHUM	57
Damascus, Syrian Arab Republic	33.42N, 36.52E, 609m	16.7	MSLP, Z500, H500, RHUM	62
Deir Ezzor, Syrian Arab Republic	35.32N, 40.15E, 212m	20.3	MSLP, ZSUR, F500, H500, RHUM	60
Hama, Syrian Arab Republic	35.12N, 36.75E, 303m	17.8	MSLP, USUR, DSUR, H500, R850	59
Kamishli, Syrian Arab Republic	37.05N, 41.22E, 455m	18.9	MSLP, USUR, ZSUR, DSUR, H500	73
Kfardane, Lebanon	34.01N, 36.03E, 1080m	15.4	MSLP, USUR, ZSUR, Z500, H500, R500, R850	50
Lattakia, Syrian Arab Republic	35.53N, 35.77E, 7m	19.5	MSLP, VSUR, Z500, H500, F850, R850	54
Palmyra, Syrian Arab Republic	34.55N, 38.30E, 404m	19.1	MSLP, DSUR, H500, R500, R850	73

Sites	Lat./ Lon./ Elev.	Mean (mm)	Precipitation predictors	E %
Aleppo, Syrian Arab Republic	36.18N, 37.20E, 384m	260	Z500, F850, R500, R850	7
Amman, Jordan	31.59N, 35.59E, 780m	238	USUR, H500, D500, F850, Z850, R500	29
Beirut, Lebanon	33.82N, 35.43E, 19m	685	USUR, F500, Z500, V850, Z850, R500, R850	12
Damascus, Syrian Arab Republic	33.42N, 36.52E, 609m	138	F500, V500, Z500, F850, R500, RHUM	12
Deir Ezzor, Syrian Arab Republic	35.32N, 40.15E, 212m	228	V500, Z500, H500, R500, R850	3
Hama, Syrian Arab Republic	35.12N, 36.75E, 303m	283	H500, F850, U850, V850, Z850, R850	25
Kamishli, Syrian Arab Republic	37.05N, 41.22E, 455m	430	USUR, V500, F850, V850, H850, R500, R850	19

(table continues on next page)

Sites	Lat./ Lon./ Elev.	Mean (°C)	Temperature predictors	E %
Kfardane, Lebanon	34.01N, 36.03E, 1080m	420	USUR, VSUR, DSUR, H500, Z850	18
Lattakia, Syrian Arab Republic	35.53N, 35.77E, 7m	793	DSUR, V500, H500, F850, U850, V850, R500	15
Palmyra, Syrian Arab Republic	34.55N, 38.30E, 404m	138	USUR, U500, H500, H850, R500, RHUM	13

Source: World Bank data.

Notes: The percentage of explained variance (E %) is shown for available data within the period 1961–2000. Annual means are based on downscaled estimates for the calibration period.

MSLP (mean sea level pressure), U* (zonal component of airflow), V* (meridianal component of airflow), F* (strength of airflow), Z* (vorticity), H* (geopotential height), R* (relative humidity), S* (specific humidity). Elevation of predictors: *SUR (near surface), *850 (at 850 hPa pressure level), *500 (at 500 hPa pressure level).

Changes in Seasonal and Annual Mean Temperature (°C) Downscaled From HadCM3 Under SRES A2 and B2 Emissions Scenarios for Selected Sites in Jordan, Lebanon, and the Syrian Arab Republic

Sites	DJF		MAM		JJA		SON		ANNUAL	
	A2	B2	A2	B2	A2	B2	A2	B2	A2	B2
2020s										
Aleppo	0.7	0.8	1.5	1.7	1.3	1.9	1.5	1.6	1.2	1.5
Amman	0.8	1.1	1.4	1.6	1.4	1.8	1.4	1.6	1.3	1.5
Beirut	0.6	0.8	0.9	1.1	0.6	0.9	0.9	1.1	0.8	0.9
Damascus	0.7	1.0	1.7	1.9	1.6	2.1	1.7	1.8	1.4	1.7
Deir Ezzor	0.7	0.9	1.4	1.8	1.1	1.6	1.4	1.5	1.2	1.4
Hama	0.6	0.7	1.4	1.7	1.3	1.9	1.4	1.5	1.2	1.4
Kamishli	0.8	1.0	1.4	1.8	1.4	2.0	1.6	1.8	1.3	1.6
Kfardane	0.8	1.0	1.5	1.8	1.2	1.7	1.2	1.3	1.2	1.5
Lattakia	0.7	0.9	1.1	1.2	0.5	0.8	1.0	1.2	0.8	1.0
Palmyra	0.7	0.9	1.5	1.8	1.5	2.1	1.6	1.7	1.3	1.6
2050s										
Aleppo	1.3	1.3	2.4	2.2	2.8	2.4	2.7	2.3	2.3	2.1
Amman	1.7	1.7	2.3	2.1	2.8	2.4	2.5	2.2	2.3	2.1
Beirut	1.2	1.2	1.5	1.4	1.3	1.1	1.7	1.4	1.4	1.3
Damascus	1.5	1.5	2.9	2.5	3.4	2.8	3.0	2.6	2.7	2.4
Deir Ezzor	1.4	1.4	2.2	2.1	2.3	1.9	2.5	2.2	2.1	1.9
Hama	1.2	1.3	2.3	2.2	2.8	2.4	2.6	2.3	2.2	2.0
Kamishli	1.6	1.6	2.3	2.2	2.9	2.5	2.9	2.5	2.4	2.2
Kfardane	1.5	1.5	2.6	2.2	2.5	2.0	2.2	1.9	2.2	1.9
Lattakia	1.5	1.5	1.8	1.6	1.2	1.1	1.9	1.6	1.6	1.5
Palmyra	1.5	1.6	2.5	2.3	3.1	2.7	2.8	2.4	2.5	2.3

(table continues on next page)

Sites	DJF		MAM		JJA		SON		ANNUAL	
2080s										
Aleppo	2.6	1.8	4.7	3.4	4.4	3.5	4.4	3.5	4.0	3.0
Amman	3.3	2.3	4.3	3.1	4.4	3.4	4.1	3.1	4.0	3.0
Beirut	2.3	1.7	2.9	2.1	2.1	1.6	2.8	2.2	2.5	1.9
Damascus	2.9	2.1	5.2	3.7	5.3	4.1	4.9	3.7	4.6	3.4
Deir Ezzor	2.7	2.0	4.7	3.2	3.6	2.8	4.2	3.3	3.8	2.8
Hama	2.3	1.7	4.6	3.3	4.4	3.5	4.4	3.4	3.9	3.0
Kamishli	3.2	2.2	5.0	3.5	4.4	3.6	4.8	3.7	4.3	3.2
Kfardane	2.9	2.1	4.9	3.5	4.0	3.1	3.6	2.8	3.8	2.9
Lattakia	2.9	2.1	3.3	2.4	1.9	1.5	3.2	2.5	2.8	2.1
Palmyra	2.9	2.1	4.9	3.5	4.9	3.9	4.6	3.6	4.3	3.2

Source: World Bank data.

Note: DJF = December, January, February; MAM = March, April, May; JJA = June, July, August; SON = September, October, November.

APPENDIX E

Changes in Seasonal and Annual Precipitation Totals (mm) Downscaled from HadCM3 Under SRES A2 and B2 Emissions Scenarios for Selected Sites in Jordan, Lebanon, and the Syrian Arab Republic

Sites	DJF		MAM		JJA		SON		ANN	
	A2	B2	A2	B2	A2	B2	A2	B2	A2	B2
2020s										
Aleppo	−4	−6	−7	−11	0	0	−9	−7	−21	−24
Amman	−17	−23	−13	−17	−1	−2	−5	−6	−36	−49
Beirut	−7	−22	−21	−28	−6	−12	−13	−14	−47	−76
Damascus	−1	−5	−4	−10	−1	−2	0	−3	−6	−20
Deir Ezzor	+2	+3	0	0	0	0	−1	+3	+1	+5
Hama	−5	−12	−4	−11	−7	−8	−5	−11	−21	−43
Kamishli	−2	−18	−17	−27	−1	−1	−3	−6	−23	−52
Kfardane	−20	−31	−18	−29	−5	−10	−15	−17	−60	−87
Lattakia	−34	−49	−33	−34	−9	−13	−38	−63	−114	−161
Palmyra	0	−1	−4	−9	0	−2	−1	−1	−4	−12
2050s										
Aleppo	−8	−4	−4	−4	0	−1	−13	−4	−26	−13
Amman	−27	−24	−17	−18	−2	−2	−12	−8	−60	−52
Beirut	−21	−16	−14	−17	−10	−4	−28	−14	−74	−52
Damascus	−5	−6	−5	−10	−3	−3	−4	−4	−17	−23
Deir Ezzor	+3	+6	+4	0	+2	+3	+2	+5	+12	+15
Hama	−13	−12	−2	−6	−14	−8	−18	−14	−47	−40
Kamishli	−16	−18	−17	−23	−2	−1	−13	−6	−48	−48
Kfardane	−40	−44	−30	−31	−13	−10	−28	−20	−113	−107
Lattakia	−82	−77	−47	−40	−19	−9	−83	−75	−234	−204
Palmyra	−2	−1	−3	−5	−1	−1	−5	−3	−11	−10

(table continues on next page)

Sites	DJF		MAM		JJA		SON		ANN	
	A2	B2	A2	B2	A2	B2	A2	B2	A2	B2
2080s										
Aleppo	−15	−3	−19	−10	−2	0	−17	−10	−54	−23
Amman	−48	−31	−36	−30	−3	−3	−16	−12	−105	−77
Beirut	−51	−29	−56	−31	−18	−14	−33	−22	−160	−96
Damascus	−9	−6	−14	−10	−4	−4	−6	−5	−34	−25
Deir Ezzor	+4	+7	−4	+3	0	+2	+3	+3	+3	+17
Hama	−30	−11	−17	−11	−19	−13	−26	−22	−93	−57
Kamishli	−40	−12	55	−39	−3	−2	−23	−10	−122	−64
Kfardane	−84	−55	−67	−48	−18	−14	−37	−28	−208	−148
Lattakia	−133	−79	−82	−63	−25	−18	−124	−99	−369	−261
Palmyra	−6	−1	−16	−12	−3	−2	−10	−6	−35	−21

Source: World Bank data.

Priority Elements of Draft Action Plan, Bekaa Valley

Response Option	Proposed Research Project	Brief Summary of Activities	Institutions	Time Table	Funding ($ US)	Expected Results & Impacts
1. Adoption of New Irrigation Technologies	Promotion of Drip Irrigation Systems using Demonstration Plots	- Disseminate LARI experience with drip irrigation through LARI-run demonstration plots, training, booklets and brochures. - Train farmers in timing and quantity of water application.	Lebanese Agricultural Research Institute (LARI), Lebanon's Ministry of Agriculture, NGOs, cooperative societies and municipalities	2 years	$800,000	- Installation of 100 ha. of demonstration drip irrigation systems. - Increased adoption of drip irrigation by farmers. - Water applications decreased by 30–40 percent per crop while increasing yields.
	Farmer Copayment System for Drip Irrigation	- Develop copayment system for drip irrigation involving cost-sharing. - Work with NGOs, cooperative societies, and municipalities to assist in farmer collaboration and communication.				
2. Pilot Project to Construct Small and Medium-scale Water Harvesting Reservoirs	Securing Non-Conventional Water Resources through Reservoir Construction	- Construct small ponds/tanks and medium-sized reservoirs for water harvesting. - Target high-quality potato production in remote areas.	Lebanese Agricultural Research Institute (LARI), Lebanon's Ministry of Agriculture, the Remote Sensing Center (National Council for Scientific Research), NGOs	2–3 years	$750,000 plus research and potato seed production costs	- Construction of 30 small water reservoirs. - Increased water availability and reduced reliance on water wells. - Increases to farmers' net incomes by 50 percent.

(table continues on next page)

Response Option	Proposed Research Project	Brief Summary of Activities	Institutions	Time Table	Funding ($ US)	Expected Results & Impacts
3. Integrated Production Management of Pests, Diseases and Plant Disorders under Climate Change	Research on Local Plant Diseases, Pests and Physiological Disorders to Support Farmer Adaptation to Climate Change	- Observe, monitor and inspect crops to identify emerging diseases and pests of key crops in Baalbeck–Hermel region. - Develop and apply Integrated Pest Management (IPM) strategies for emerging diseases and pests. - Promote Best Management Practices (BMPs) in demonstration fields.	LARI Department of Plant Protection, Department of Irrigation, local farmers and farmer organizations	2–3 years	$310,000	- Identification of emerging plant diseases, pests, and physiological disorders and their solutions in the Baalbeck-Hermel region. - Increased adoption of IPM practices and BMPs by farmers. - Potential yield increases of 5–10 percent and reduction of pesticide spraying.
4. Production and Dissemination of Crops and Plants Adapted to Climate Change	Production, Certification, and Delivery of Local Climate Change-Adapted Plant Species and Varieties	- Select locally economically important rootstocks and varieties known for adaptability to climate change	Machatel Loubnan Nursery Association, private nurseries, farmers in Baalbeck-Hermel region, Lebanese Agricultural Research Institute (Biotechnology Department, Plant Protection Department), and Ministry of Agriculture's Plant Certification Department	2–8 years	$700,000	- Production of 100,000 plants distributed to farmers covering 625 ha. - Yield, quality, and shelf life improvements. - Reduction of irrigation water use.
	Production, Certification, and Delivery of International Climate Change-Adapted Plant Species and Varieties	- Introduce internationally important rootstocks and varieties known for drought tolerance for assessment of tolerance to climate change impacts		2 years		
5. Capacity Building for Climate Change Adaptation	Establishment of a National Climate Change Authority	- Develop monitoring system to recommend adaptation and mitigation measures. - Develop database to support farmer adaptation strategies.	Lebanese Agriculture Research Institute (Departments of Plant Biotechnology, Irrigation and Agrometeorology, Plant Protection, Plant Breeding, and the Central Laboratory),	2–3 years	$560,000	- Improvements to farmers' ability to address climate change impacts while improving productivity and quality of crops.

(table continues on next page)

Response Option	Proposed Research Project	Brief Summary of Activities	Institutions	Time Table	Funding ($ US)	Expected Results & Impacts
		- Develop early warning systems to provide daily weather predictions, seasonal forecasts.	Ministry of Agriculture (Extension Department), INRA (Institut National de la Recherche Agronomique-France), CIHEAM–Bari (Centre International des Hautes Etudes Agronomiques Mediterranéennes), CIMA Foundation (International Centre For Environmental Monitoring, Italy), and Saint Joseph University			
	Improving Skills, Knowledge, Research, Collaborations and Connections at the LARI	- Improve quality and relevance of LARI-conducted research, technical knowledge, and skills. - Strengthen collaborations between LARI and agricultural faculty at technical and university institutions and other national and regional institutions. - Facilitate communication between LARI and farmers through joint training sessions and workshops.				- Lebanon-focused research on climate change effects in agriculture. - Improved awareness, expertise, and communication between scientists, government, and farmers on climate change and adaptation measures.
	Integrating Climate Change Adaptation Plans into National Strategies and Activities	- Screen and revise the national development plan and rural development strategies. - Embed climate adaptation plans in Ministry of Agriculture activities.				- Improved integration of climate change adaptation strategies across national agencies and organizations.

(table continues on next page)

Response Option	Proposed Research Project	Brief Summary of Activities	Institutions	Time Table	Funding ($ US)	Expected Results & Impacts
6. Evaluation and Maintenance of Genetic Diversity of Wild Species and Local Varieties Adapted to Climatic Change	Conservation and Development of Key Crops and their Genetic Diversity	- Survey, collect, define, and assess the genetic diversity of the local varieties and wild species of key crops: wheat, barley, *Prunus*, fig, and caper. - Conserve distinct varieties and species both *in-situ* (on-farm) and *ex-situ* (gene bank). - Evaluate drought tolerance of local varieties by using in-vitro techniques.	Lebanese Agricultural Research Institute (Depts. of Plant Biotechnology, Irrigation, Plant Protection, and Plant Breeding), ICARDA (International Center for Agricultural Research in Dry Areas), CNRS (National Council for Scientific Research), and ACSAD (Arab Center for the Studies of Arid Zones and Dry Lands)	10 years	$1,450,000	- Conservation of genetic diversity to address future climatic changes.
	Introduction of Crops for Breeding Programs and Cultivation	- Introduce wild and local varieties in plant improvement programs on-farm and through field trials with farmers. - Adapt and produce selected varieties as certified drought-tolerant material by LARI. - Distribute certified varieties to farmers at low prices.				- Improved, more productive and more diversified crop varieties, resulting in higher production and increased farmers' incomes.

Source: World Bank data.

Priority Elements of Draft Action Plan, Jordan River Valley

Response Option	Proposed Research Project	Brief Summary of Activities	Institution	Time table	Funding ($)	Expected Results & Impacts
1. Improve Farm Production Systems and Productivity	Adapting Crops and Cropping Patterns to Climate Change	- Evaluate and introduce new crop varieties that are resilient to anticipated climatic changes. - Evaluate current cropping system constraints and potentials. - Identify and introduce alternative cropping patterns and cultural practices for climate change resiliency.	NCARE Departments of Field Crops, Horticulture, Plant Protection, Olive Trees, Water, Soil & Environment, and the Drought Monitoring Unit, Ministry of Agriculture, and Jordan Meteorological Department.	5 years	$5,000,000	- Evaluation of current constraints. - Introduction of new crops/ varieties and management practices better suited for climatic changes.
	Conservation Agriculture and Wheat Landraces for Climate Change Adaptation	- Increase seed production and collect resilient wheat landraces. - Establish on-farm conservation agriculture fields. - Disseminate and promote conservation agriculture practices.	NCARE and University of Jordan.	3 years	4,000,000	- Identification of wheat landraces and genotypes adapted to no-till system. - Increased adoption of conservation agriculture.

(table continues on next page)

Response Option	Proposed Research Project	Brief Summary of Activities	Institution	Time table	Funding ($)	Expected Results & Impacts
	Diversity and Conservation of Barley Genetic Resources to Mitigate Climate Change Impacts	- Collect and conserve wild barley genetic resources. - Use these resources to address climatic constraints to barley production such as high heat or low moisture conditions.	International Centre for Agricultural Research in Dry Area (ICARDA), NCARE, Ministry of Education (MOE), Agricultural Schools, and NGOs	5 years	300,000	- Identification, conservation, and use of promising wild barley accessions to improve barley production under climatic constraints
	Conservation of Medicinal Herbal Plants *in situ* and *ex situ* and their Sustainable Utilization	- Conserve genetic resources of medicinal plants *in situ* and *ex situ*. - Characterize and use genetic resources of medicinal plants. - Support the sustainable use of medicinal plants to diversify farmers' production systems.	Biodiversity program/NCARE and GIS Unit/NCARE	5 years	2,000,000	- Conservation, characterization, and use of major medicinal plants.
2. Improve On-farm Water Use Efficiency and Integrated Water Resources Management	Promotion of Rainwater Harvesting	- Promote and implement rainwater harvesting technologies and practices.	NCARE, Ministry of Agriculture, The University of Jordan, Jordan University of Science and Technology, Badia Research and Development Center, and Royal Geographic Center	3 years	400,000	- Dissemination of rainwater harvesting systems.
	Advanced Wastewater Treatment Technology and Reuse	- Design and operate improved wastewater treatment systems for selected fodder and cut flower crops.	National & regional institutions	3 years	97,000	- Economically and technically feasible operation of wastewater treatment systems
	Reuse of Greywater in Homes and Farming	- Analyze the economic feasibility of greywater treatment and reuse for irrigation. - Evaluate the environmental impacts of using treated grey water for irrigation.	NCARE and rural communities	Long term	100,000	- Sustainable, economical, and improved use of scarce water resources in agricultural production.

(table continues on next page)

Response Option	Proposed Research Project	Brief Summary of Activities	Institution	Time table	Funding ($)	Expected Results & Impacts
	Deficit Irrigation for Improving Water Productivity of Vegetable Crops	- Evaluate the economic and environmental potential for deficit irrigation for vegetable crops.	Water Management and Environment Department/ NCARE	3 years	176,000	- Determined feasibility of deficit irrigation for vegetable production.
3. Improve Livestock and Rangeland Systems	Breeding for Climate Resilient Small Ruminant Livestock	- Target genetic characteristics of endemic livestock through breeding for dry climate tolerance.	Integrated Livestock and Rangeland Department/ NCARE	Long-term	$100,000	- Development of heat-tolerant breeds. - Sustained production and rural incomes.
	Rangeland and Livestock Management for Climate Change Resiliency	- Improve animal nutrition and other management practices - Improve monitoring of rangeland conditions. - Rehabilitate degraded rangeland and maintain rangeland biodiversity. - Increase awareness of land-use impacts among farmers and rural communities.	Integrated Livestock and Rangeland Directorate/ NCARE and Biodiversity and Medicinal Plants Directorate/ NCARE	5 years	1,640,000	- Improved rangeland protection and conservation. - Increased awareness of climate change impacts and adaptation strategies in rural households.
4. Building National Capacity	Capacity Building for Climate Change Adaptation	- Establish an integrated regional database to track climate changes in the Jordan River Valley. - Increase researcher capacity, climate change relevance, and collaborations with farmers at NCARE. - Develop a National Agricultural Climate Change Strategy. - Co-ordinate climate change mitigation and adaptation and natural resource management across public institutions.	NCARE, farmers, and related agricultural, water and environmental institutions	Long-term target	1,000,000	- Development of researcher and farmers' capacity to respond to climate changes - Development of infrastructure to address new challenges posed by climate change and to enhance the agricultural system

(table continues on next page)

Response Option	Proposed Research Project	Brief Summary of Activities	Institution	Time table	Funding ($)	Expected Results & Impacts
5. Reduce Risks of Agricultural Pests and Diseases	Reduce risk of agricultural diseases and pests	- Study the impact of climate on population dynamics of plant diseases and pests. - Determine suitable cultural practices to reduce infestations, including Integrated Pest Management (IPM) practices	Land Protection Research Department/ NCARE	5 years	250,000	- Reduced frequency and severity of outbreaks or infestations of disease, pests and weeds. - Improvement to farm household incomes.
6. Reinforce Early Warning System for Drought	Reinforce Early Warning System for Drought	- Evaluate alternative remotely sensed indices for drought on a real-time basis, and their relevance for Jordan. - Perform field studies of drought effects as correlated to remotely sensed indices. - Perform socio-economic study for selected areas to identify potential economic damages of drought.	Drought Monitoring Unit/NCARE, Jordan Meteorological Department, Ministry of Water and Irrigation Water, Soil and Environment, Field Crop Dept., Horticulture Dept., Integrated Livestock and Rangeland, Socio Economic Studies/NCARE	3 years	400,000	- Development of more effective methods and indicators to minimize the impacts of drought on people and agriculture. - Increased farmer income and sustainability due to improved drought response
7. Reform Land-Use Laws and Implement Sustainable Land-Use	Review Land-use Laws for Enhancement of Sustainable Land-use	- Review existing laws and propose changes to enhance sustainable land-use practices. - Improve mapping of soils, land-use status and desertification risk.	Ministry of Agriculture, NCARE, Ministry of Water and Irrigation, Ministry of Environment, Ministry of Municipal Affairs, Ministry of Justice, and Department of Land and Surveys/ Ministry of Finance, Prime Minister	2 years	200,000	- Improved and more sustainable land-use management and planning.
	Targeting Soil and Water Conservation to Land-use Characteristics	- Study and analyze land-use and land cover, land ownership, parcel size, land fragmentation and land-use suitability, and identify appropriate soil water conservation techniques.	NCARE, Ministry of Agriculture, Ministry of Water and Irrigation, Ministry of Environment, Jordanian universities, Faculty of Agriculture, and stakeholders at different governorates	3 years	1,662,000	- Improved conservation as targeted to different land-uses.

(table continues on next page)

Response Option	Proposed Research Project	Brief Summary of Activities	Institution	Time table	Funding ($)	Expected Results & Impacts
8. Activation of Agricultural Risk Management-Fund	Activate the Agricultural Risk Management Fund	- Implement the pre-existing Agricultural Risk Management Fund to help farmers better cope with risk. - Identify legislation, regulations, tools, financing and infrastructure needed for the Fund to function effectively. - Implement Fund if feasible.	Ministry of Agriculture, Agricultural Risk Management Fund, farmers, and private insurance companies	Long-term	5,000,000	- Identification of needs for effective functioning of an Agricultural Risk Management Fund. - Possible fund implementation.

Source: World Bank data.